Fruit Is Good For Me

Count, Color, Draw

written and illustrated by

Annett M. Prince

Fruit Is Good For Me

Color, Color, Draw

Written and illustrated by

Annett M. Prince

Published by Astute Motivators & Innovators, LLC

ISBN: 978-1-962588-00-3 (Hardcover)
ISBN: 978-1-962588-01-0 (Paperback)
ISBN: 978-1-962588-05-8 (eBook)

Visit us on the web at www.successapplications.com
Email: kids@successaplications.com

A K.I.D.S (Kids Inviting Daily Success) Publication

Fruit

Is Good For Me

Count, Color, Draw

How many?

How many apples can be found in the picture on the next page? Count them to find out.

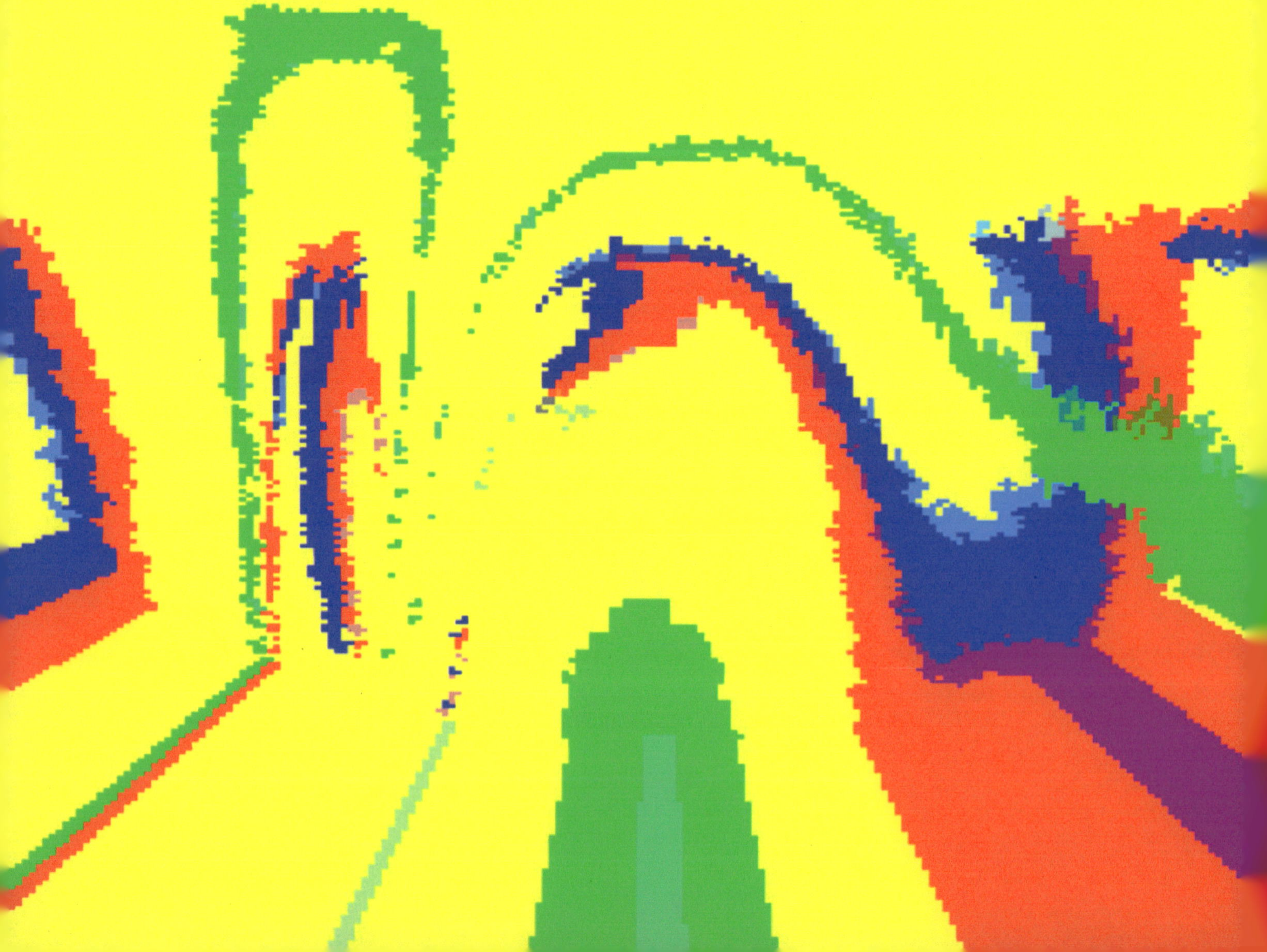

How many?

How many red apples are in the puzzle on the next page? Count them to find out.

How many?

Count the number of green apples in the picture on the next page?

Name This Fruit

On the next page, write the letters to complete the spelling of the fruit shown below.

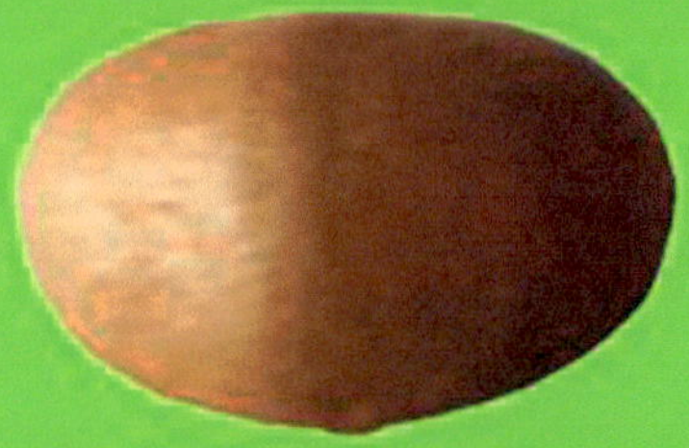

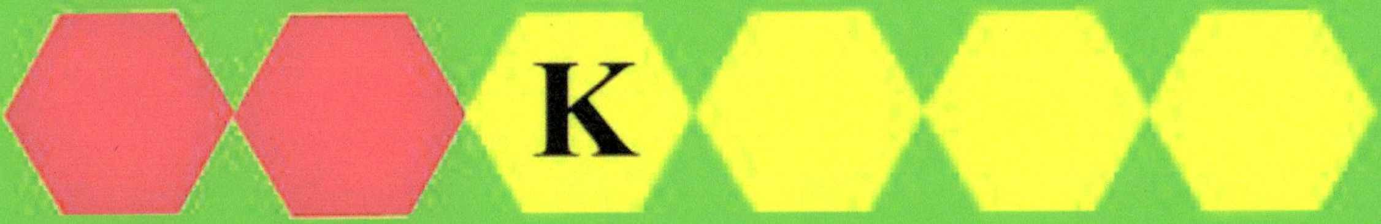

The first letter is written for you.

Name The Fruit

On the next page, write the letters to complete the spelling of the fruit shown below.

The first letter is written for you.

o

Name This Fruit

On the next page, write the letters to complete the spelling of the fruit shown below.

The first letter is written for you.

Name This Fruit

On the next page, write the letters to complete the spelling of the fruit shown below.

The first letter is written for you.

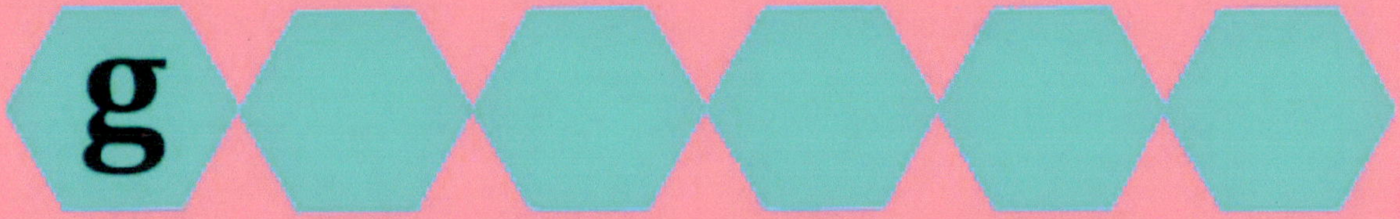

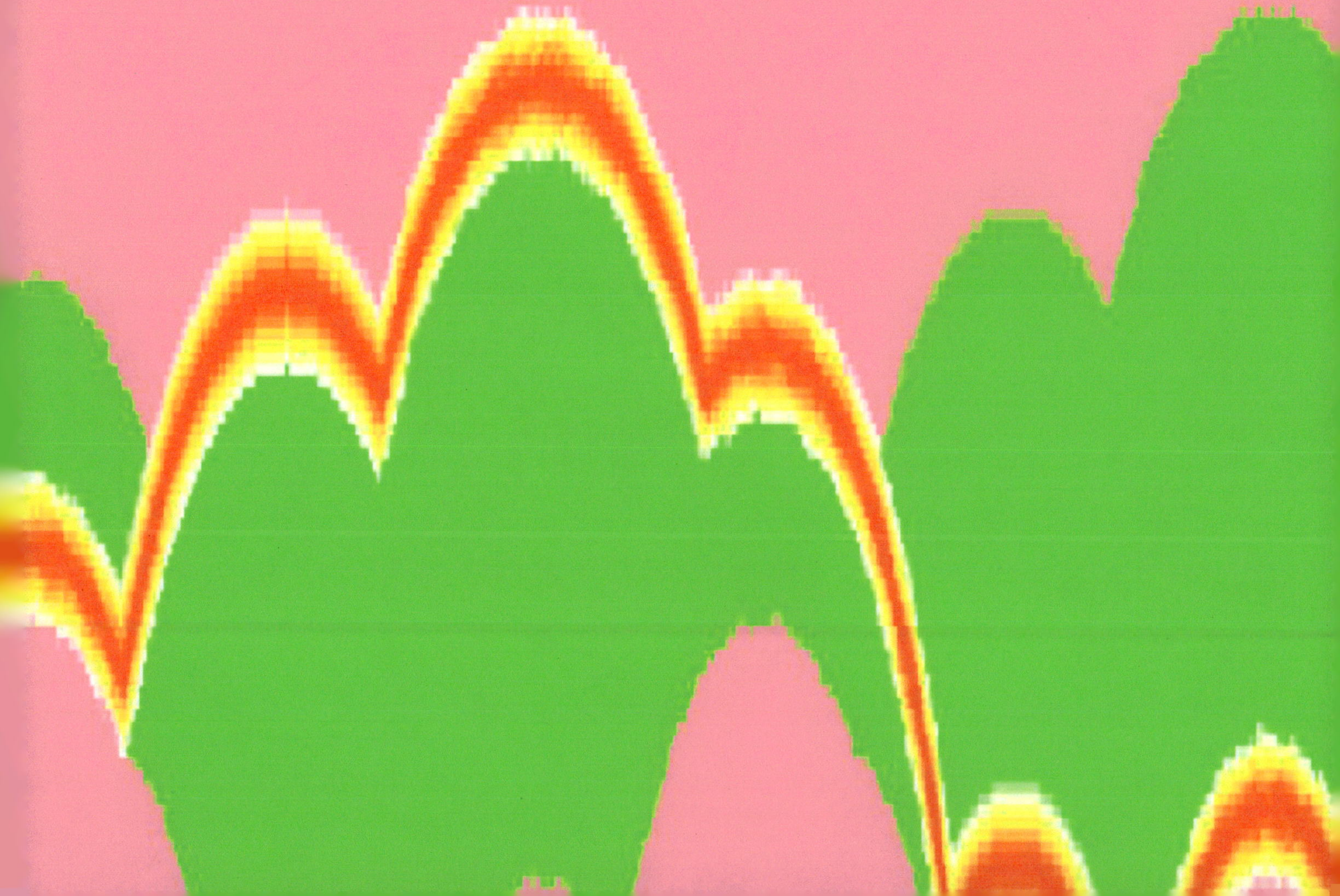

Name This Fruit

On the next page, write the letters to complete the spelling of the fruit shown below.

The first letter is written for you.

Name This Fruit

On the next page, write the letters to complete the spelling of the fruit shown below.

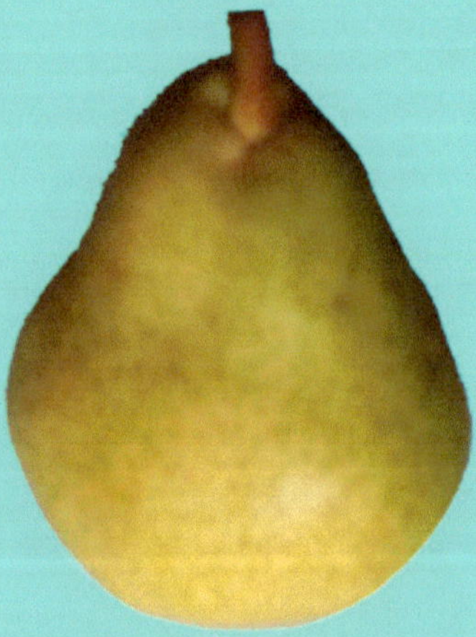

The first letter is written for you.

Count the Strawberries

On the next page, count the number of strawberries that are partially hidden in the puzzle. Then, write the number in the space provided.

There are _____ strawberries in this puzzle.

Count the Bananas

On the next page, count the number of bananas. Then, write the number in the space below.

There are ______ bananas on the next page.

Can You Tell?

On the next page, can you tell how many bananas it took to create the letter S?

S

Can You Tell?

On the next page, can you tell how many grapefruit it took to create the letter B?

B

Tangerines

The tangerines below are placed to form the letter T.

Count the tangerines to see how many were used to form the letter T?

Cross Outs

On the next page, find all the letters that are not in the word apple.

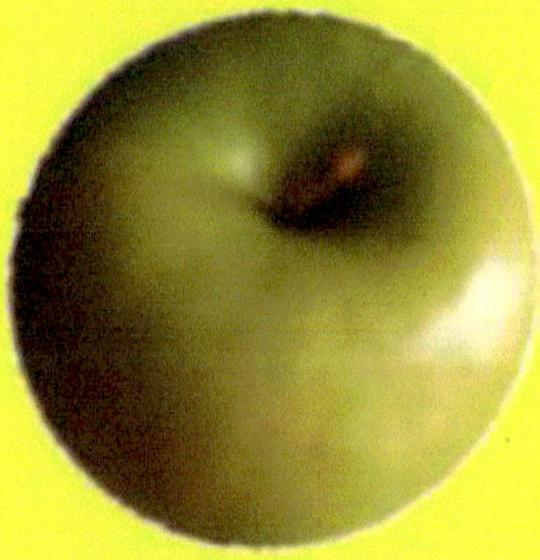

S	K	O	M	K	C	R	N
C	D	G	V	O	C	S	I
M	A	P	P	L	E	N	S
G	H	C	I	K	T	O	G
T	P	V	K	M	H	K	M
H	M	O	G	O	I	B	H

Find Out

On the next page, find all the letters that are not in the word pear.

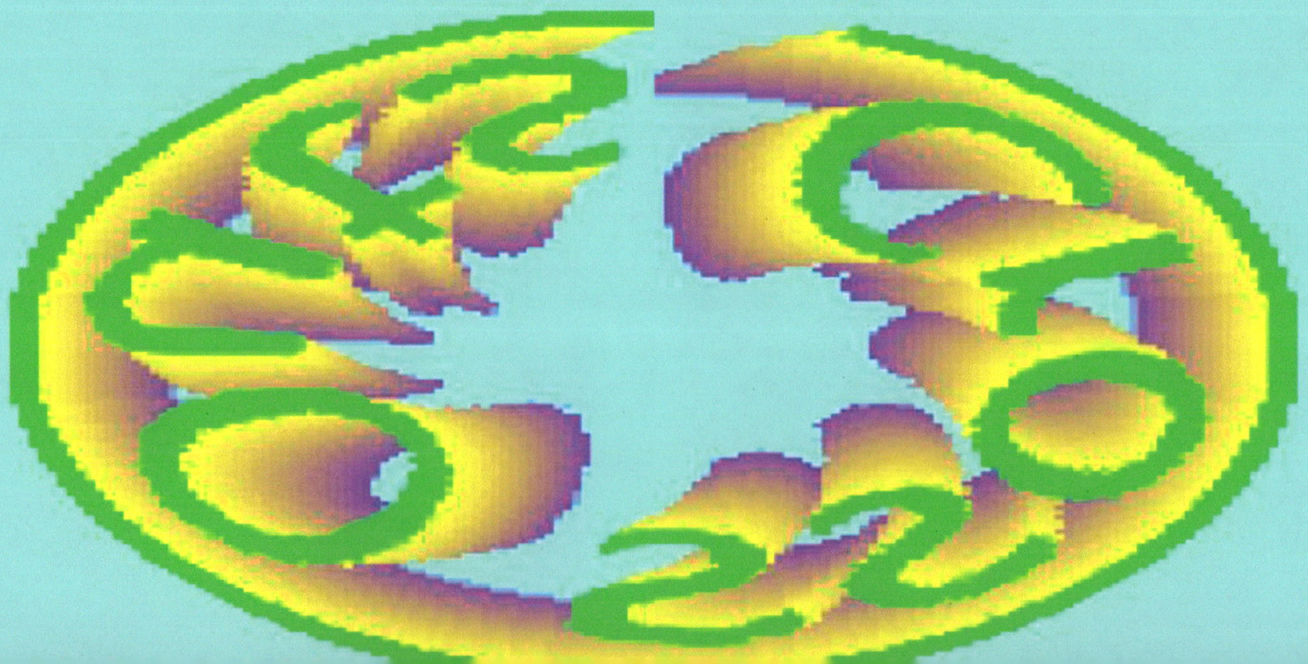

S	K	O	M	K	C	Q	N
C	D	G	V	O	C	S	I
M	L	P	E	A	R	N	S
G	H	C	I	K	T	O	G
T	Q	V	K	M	H	K	M
H	M	O	G	O	I	B	H

Draw

Draw a fruit that is yellow or orange in color.

Draw

Draw a fruit that is red or green in color.

A is for Apple

The word apple begins with the letter A.

On the next page, count the number of apples that are shown in the letter A.

On the next page, draw lines to connect the apples, to completely form the letter A.

Color the letters on the next page, with your favorite colors.

Apple Connect and Color

Draw a line to connect the apples on the next page, and color inside the circle of green apples.

Peach

The word peach begins with the letter

P

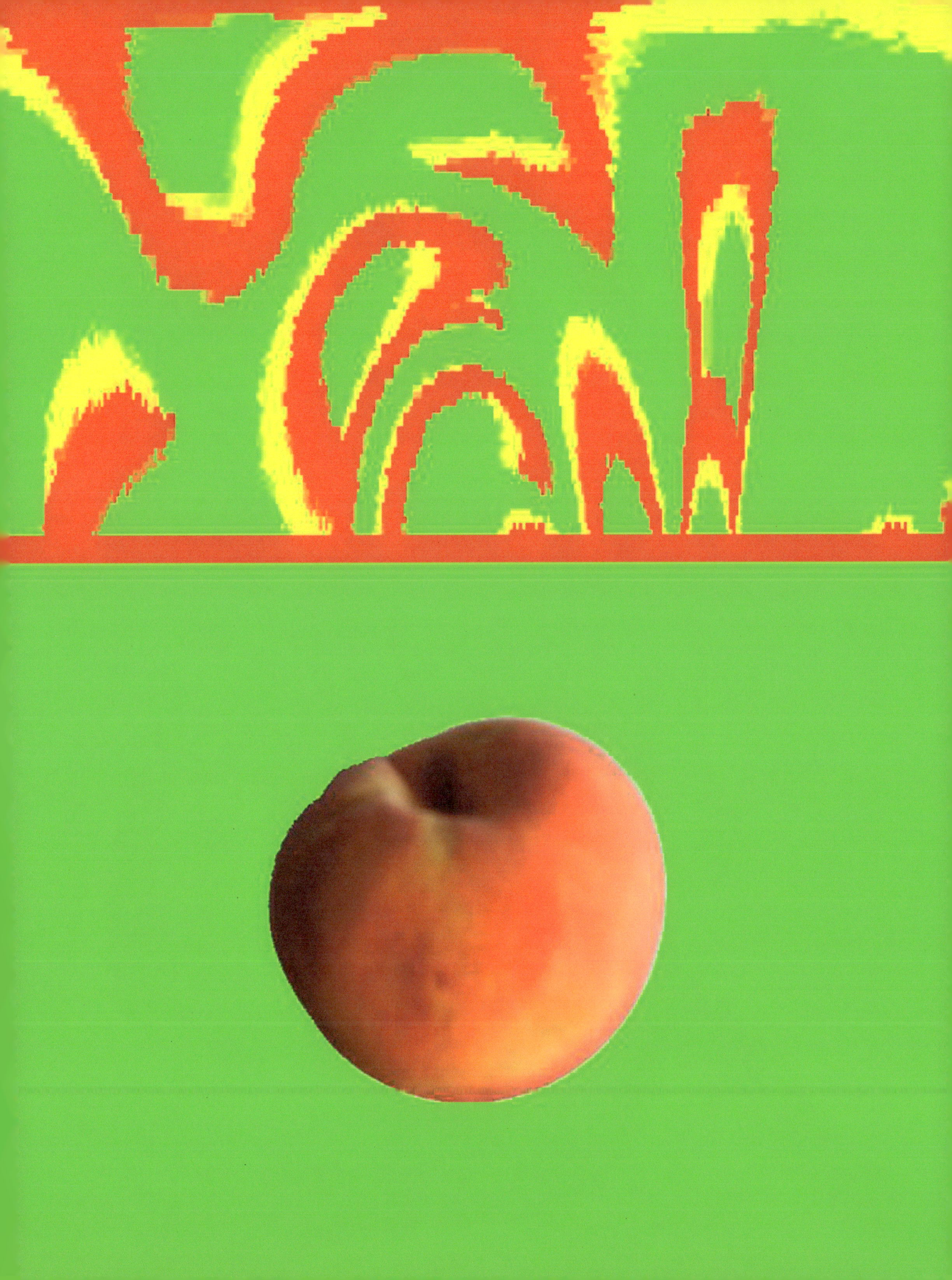

Peaches

On the next page, connect the peaches to completely form the letter P.

Count the number of peaches that form the letter P, the first letter of the word peach.

Count the number of peaches on this second letter of the word peach.

Count the number of peaches on this third letter of the word peach.

Count the number of peaches on this fourth letter of the word peach.

Count the number of peaches on this fIfth and last letter of the word peach.

How many peaches do you see on the letters below?

The name of this fruit begins with the letter

G

G

is for Grapes

Connect the dots on the next page to form the letter G.

The name of this fruit begins with the letter O. On the next page connect the oranges to completely form the letter O.

On the next page, draw oranges to complete the circle to form the letter O.

Take Away

If two of the pears on the next page were taken away, how many would be left?

Take Away

If three of the pears are taken away from the next page, how many would be left?

Add & Count

On the next page, add the number of pears below the circle, to the number of pears in the circle. How many pears are there?

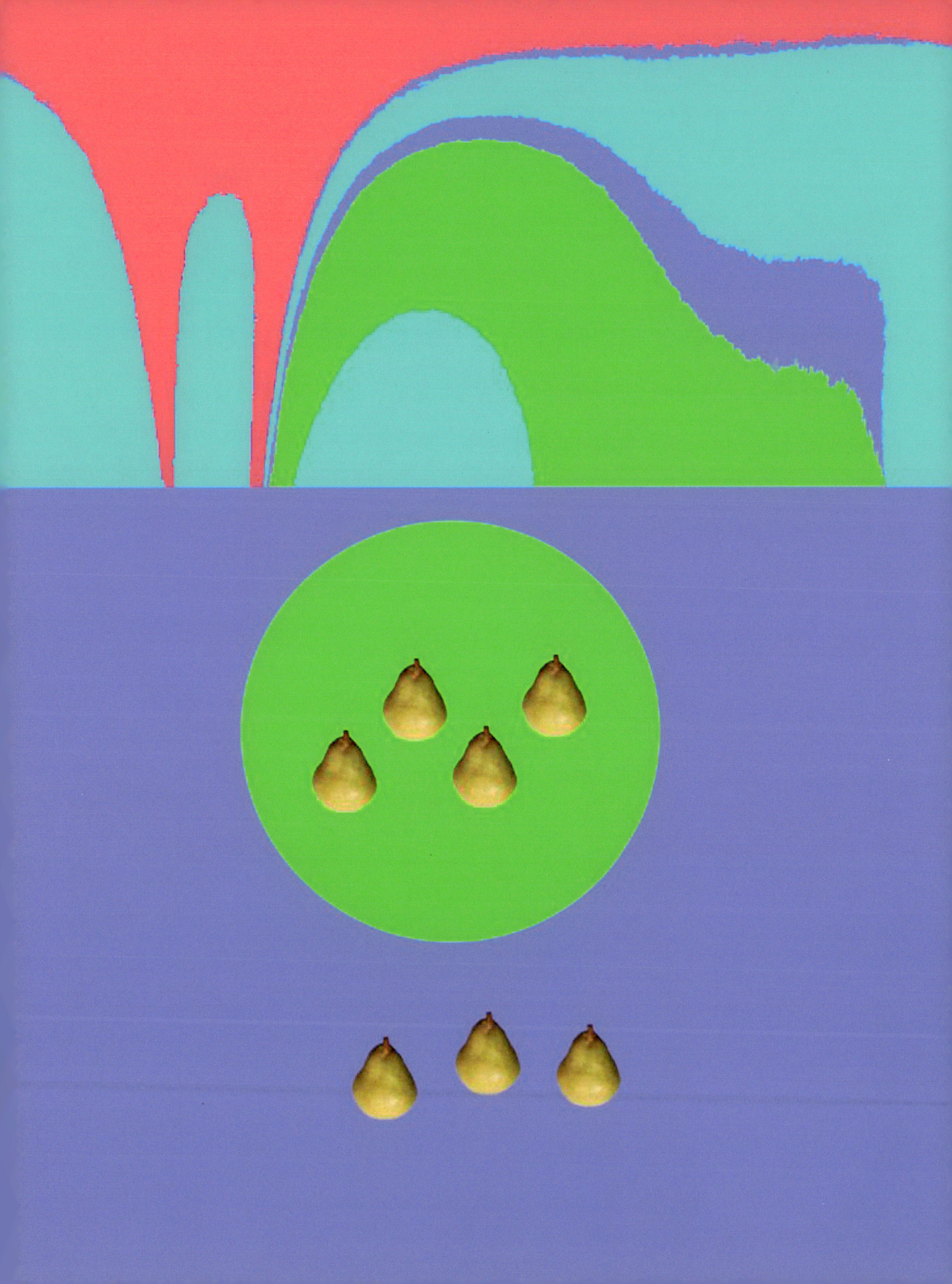

Add & Count

On the next page, count the number of pears in each group, then add the numbers together, to find out the total number of pears.

Trace Your Way Out

How quickly can you get to the apple on the next page? Using the numbers as a guide, draw a line to connect all the shapes of the same color to its highest number. See which color will lead you to the apple.

39
20
19
18
38
12
29
17
28
27
19
20
21
16
37
26
8
13
22
15
36
25
14
23
35
14
7
24
34
15
24
13
23
6
25
33
12
22
16
26
32
11
21
10
27
18
19
31
17
28
29
30
20

What I Learned

From

Fruit Is Good For Me

Count, Color, Draw

Annett M. Prince

Annett M. Prince has written and illustrated four other childrens's books, as seen below:

The Vegetable Band, I AM Children - Teen Motivational Activity Book, Circle Square, Radiance The Little Star

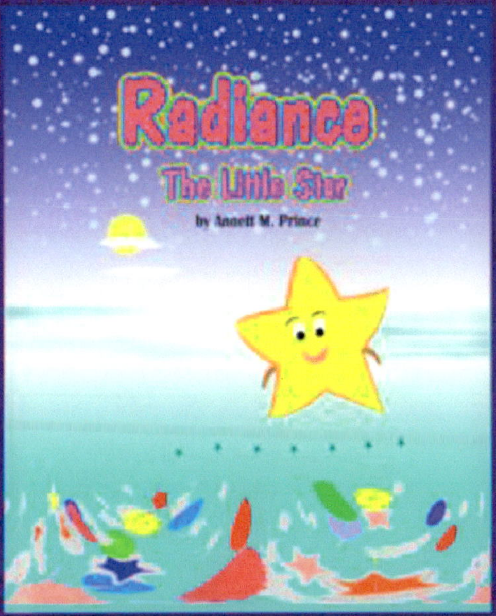

A Word From The Author

Your generosity is always greatly appreciated. Become a part of our book drive initiative to get books into the hands of children who need them.

You may give books directly to children that you know, or present books to your local school, library or any non profit, who will distribute them where it is vital and necessary.

Whether you contribute books that are authored by me or any other author, you can help to shape a young life, in a positive way.

Thank you for being a partner in this effort.

Made in the USA
Monee, IL
05 November 2026

67386671R00048

ANDREAS GÜTTLER

High accuracy determination of skin friction differences in an air channel flow based on pressure drop measurements

SCHRIFTENREIHE DES INSTITUTS FÜR STRÖMUNGSMECHANIK
KARLSRUHER INSTITUT FÜR TECHNOLOGIE (KIT)

BAND 2

High accuracy determination of skin friction differences in an air channel flow based on pressure drop measurements

BY

ANDREAS GÜTTLER

KIT Scientific Publishing

Dissertation, Karlsruher Institut für Technologie (KIT)
Fakultät für Maschinenbau, 2015

Tag der mündlichen Prüfung: 18. Mai 2015
Referenten: Prof. Dr.-Ing. B. Frohnapfel
Prof. Dr.-Ing. C. Tropea

Impressum

Scientific Publishing

Karlsruher Institut für Technologie (KIT)
KIT Scientific Publishing
Straße am Forum 2
D-76131 Karlsruhe

www.ksp.kit.edu

Print on Demand 2017 – Gedruckt auf FSC-zertifiziertem Papier

ISSN 2199-8868
ISBN 978-3-7315-0502-0
DOI 10.5445/KSP/1000053260

Abstract

The present work delivers a versatile, accurate and reliable wind tunnel facility that is employed to test different flow control techniques for drag reduction in precisely adjustable test conditions. A blower wind tunnel with a channel flow test section is developed, capable of resolving changes in the skin friction drag as small as 0.4% thanks to the accurate measurement of both pressure drop throughout the channel and volumetric flow rate. This high accuracy is verified through a detailed uncertainty analysis and is proven by means of thoroughly conducted measurements of the smooth reference test section. A comprehensive comparison with literature data reveals the correctness of the measurements under laminar and turbulent flow conditions. The capabilities of the experimental set-up are exploited to quantify the potential skin friction reduction achievable by passive riblets and micro grooved surfaces, as well as by active spanwise oscillating walls, each of which have been applied in the present work. The influence of different measurement strategies on the measurement accuracy is elaborated in detail and extensively discussed.

The results of the riblet experiments are in excellent agreement with literature, where the high accuracy of measurements allowed to prove the drag reduction over riblet surfaces to be independent of the streamwise length. Remarkably, the present work is the first experimental contribution in air with comparable accuracy to that of the most accurate studies in liquids. In contrast to existing literature, a careful analysis of the micro groove experiments uncovered that no effect on turbulent drag is achievable with such devices. This contradiction demonstrates the sensitivity of measurement accuracy and uncertainty estimation to the drawn conclusions. Furthermore, the capabilities of the present facility made possible to accurately quantify the small turbulent drag reduction obtained by spanwise wall oscillations implemented with Dielectric Elastomer Actuators (DEA).

Finally, a new correction factor is proposed in so as to account for three dimensional effects, such as finite channel width and corresponding flow development issues. It is hypothesized that this factor significantly improves the comparability of drag reduction results from numerical channel flow simulations and experimental studies of varying test-section aspect ratio.

Kurzfassung

Die vorliegende experimentelle Arbeit liefert eine vielseitige, präzise und zuverlässige Windkanalanlage zur Untersuchung verschiedener Techniken der Strömungskontrolle zur Reibungsminderung unter gleichen Testbedingungen. Ein Blaswindkanal mit einer Kanalmessstrecke wird entwickelt, dem genauste Messtechnik zur Bestimmung des Druckabfalls entlang der Messstrecke sowie des Volumenstroms erlaubt, feinste Wandreibungsdifferenzen von 0.4% aufzulösen. Diese hohe Genauigkeit wird verifiziert durch eine detailierte Fehleranalyse und anhand sorgfältig durchgeführter Messungen der glatten Referenzmessstrecke bestätigt. Ein umfangreicher Vergleich mit Literaturdaten belegt die Richtigkeit der Messergebnisse unter laminaren und turbulenten Strömungsbedingungen. Die Fähigkeiten des experimentellen Aufbaus werden genutzt, um Reibungsminderung verschiedener Techniken zur Strömungskontrolle zu quantifizieren, durch passive Riblet- und Microgroovesoberflächen sowie mittels aktiver oszillierender Wände, die alle in dieser Arbeit zur Anwendung kamen. Die Auswirkung verschiedener Messstrategien auf die Messgenauigkeit wird im Detail erarbeitet und ausführlich diskutiert.

Die Ergebnisse der Ribletexperimente zeigen eine exzellente Übereinstimmung mit Literaturwerten, wobei die hohe Messgenauigkeit die Bestätigung der Unabhängigkeit der Reibungsreduktion über Ribletflächen von der stromweitigen Position erlaubt. Bemerkenswerter Weise gelingt mit der vorliegenden Arbeit der erste experimentelle Beitrag in einer Luftströmung, der eine vergleichbare Messgenauigkeit wie die genausten Untersuchungen in Flüssigkeiten erzielt. Im Gegensatz zur existierenden Literatur deckt eine sorgfältige Analyse der Microgrooves auf, dass mit diesen keine Beeinflußung der turbulenten Reibung erzielt wird. Dieser Gegensatz zeigt die Sensibilität der Messgenauigkeit und Fehlerabschätzung auf die gezogenen Schlüsse. Außerdem ermöglicht die Anlage die geringe turbulente Reibungsminderung, die mit spannweitig oszillierenden Wänden, die mittels dielektrischer Elastomeraktuatoren (DEA) umgesetzt wurde, genau zu quantifizieren.

Darüber hinaus wird ein Korrekturfaktor vorgeschlagen, um dreidimensionale Effekte, wie begrenzte die Kanalbreite und die daraus resultierende Ausbildung der Kanalströmung zu berücksichtigen. Es wird angenommen, dass dieser Faktor zu einer deutlichen Verbesserung der Vergleichbarkeit der Reibungsminderungsergebnisse von numerischen Untersuchungen von Kanalströmungen zu experimentellen in Abhängigkeit des Seitenverhältnisses des Kanals führt.

Contents

1 Introduction

Interactions between fluids and surfaces play an essential role in many natural and industrial flows. In almost all technical flow processes skin friction at solid walls creates undesirable energy losses. During the last decades technological advances inspired the investigation of strategies aimed at manipulating the characteristics of fluid flows to minimize energy losses. Due to the potential energy savings, flow control nowadays represents a major topic of modern fluid mechanics (Gad-el Hak, 2000). One main goal of flow control is the reduction of skin friction in internal and external flows. Many strategies have been developed to achieve this goal based on the existence of two fundamental flow states:

Already in his pioneering work Osborne Reynolds (1883) observed two different forms of flow in a pipe, which he described according to the complexity of their movement as "direct" and "sinuous". The occurrence of either of these flow states, nowadays called laminar and turbulent respectively, depends on a dimensionless quantity first proposed by Stokes (1851), which has later been named after Osborne Reynolds. This so-called Reynolds number Re combines the flow velocity U, the kinematic viscosity ν of the fluid and a characteristic length L - which in case of a pipe flow is the diameter D - as follows:

$$Re = \frac{UL}{\nu}. \tag{1.1}$$

Reynolds found out that above a critical value of this number the flow transitioned from the laminar to the turbulent state and that the latter state is characterized by significantly higher wall friction: "The resistance is generally proportional to the square of the velocity (turbulent) [$\cdots$] and proportional to the velocity", if it is laminar (Reynolds, 1883).

Reynolds observed that the higher momentum exchange associated with turbulent flow causes fluid elements to "eddy about in sinuous paths", while in the laminar state they move "along direct lines", which explains the higher drag in the former. This observation constitutes the rationale of scientific research in the field of drag reduction: the manipulation of the friction intensive turbulent flow state. Two strategies of flow control in the field of drag reduction are deduced:

⇒ the sustainment of a laminar flow state at higher values of Re, for which the flow would be otherwise naturally turbulent;

⇒ a modification of the turbulent regime aimed to reduce the excess turbulent skin-friction drag.

Based on these strategies various approaches to reduce the friction drag have been developed. A differentiation of the methods can be made by classifying them into active - those consuming energy - and passive - those not utilizing external energy sources. While active techniques - mainly in interaction with complex feedback control systems - promise applicability in wide velocity ranges and high amounts of savings, methods based on passive principle score with efficiency and simplicity in operation. In spite of the large number of research projects devoted to the subject, only few flow control strategies developed to a state enabling application in laboratory experiments. In the following promising approaches for passive and active flow control methods are exemplarily presented.

One of the most famous examples for a passive method are riblets. These shark-skin like structures have been studied for more than three decades and their capabilities in reducing drag in a turbulent flow have been demonstrated (Bechert *et al.*, 1997; Lee & Lee, 2001; Walsh, 1982). The standard type of two-dimensional fin structures has been extensively studied experimentally. The effectiveness of riblets strongly depends on Re with a relatively small drag reduction of typically 4-10%.

The most successful application of skin friction drag reduction is also represented by a passive method (Toms, 1948): the addition of foreign substances to fluid flows, for example polymers with a long-chain structure. Despite a very low concentration of few ppm the polymers have shown the outstanding reduction, up to 80%, in turbulent flows (Lumley & Kobu, 1985). Such additives are used in the Transalaska pipeline for oil transport since 1979.

Micro grooves are a recently proposed passive technique suggested for turbulent drag reduction, that also showed promising results in transition delay (Jovanović *et al.*, 2010a). These streamwise aligned grooves are of very small dimensions -significantly smaller than riblets- and their fabrication is a problem to be solved. Groove structures were found to yield a high amount of skin friction drag reduction but only for narrow range of *Re* (Frohnapfel *et al.*, 2007a).

A comprehensive review on the actuator design for active flow control is provided by Cattafesta & Sheplak (2011). An example for an active flow control aimed at delaying laminar to turbulent transition, which has been in focus in the recent years, is boundary layer stabilization with plasma actuators. Its working principle is based on ionization and acceleration of air molecules, where the major advantage is the fast response time and the absence of any moving parts. A small wall jet is generated, which can be used to import momentum into the boundary layer (Duchmann *et al.*, 2013) or damp distortions, which trigger the transition process (Kurz *et al.*, 2013).

Spanwise wall forcing is an active, predetermined flow control technique for turbulent skin-friction drag reduction which can achieve high-values of drag reduction. It has been intensively studied numerically (Quadrio & Ricco, 2004) and has been given experimental verification (see, for example Ricco & Wu (2004)). Its simplest implementation is the oscillating wall, in which a portion of the wall wetted by the fluid exhibits in-plane oscillations perpendicular to the mean flow direction. In spite of its simplicity, designing a device capable of mimicking this movement, which contemporaneously meets the space- and timescales of wall turbulence and is energetically efficient, is a very challenging task.

The few flow control strategies, which developed to a state enabling application in laboratory experiments, usually achieved a skin-friction drag reduction of less than 10%. At the fairly low values of the Reynolds number, for which laboratory implementation of turbulent flow control is possible, skin-friction forces can be very low. The relatively small variation of drag created by the control is therefore very difficult to detect. Additionally many parameters influencing the flow properties need to be kept constant to ensure reproducible results, e.g. the arrangement of set-up, the fluid parameters and the environmental influence, etc.. The different research objectives comprising laminar, transitional or turbulent flow for the variety of studies led to different designs of the corresponding experimental facilities.

In summary, the boundary conditions and uncertainties of measurement differ to such an extent, that a comparative evaluation of the methods in respect to the amount of drag reduction or efficiency is rarely possible. The major goal of the present work, is therefore is to provide an experimental facility for high accurate quantification of various methods of flow control under identical reproducibility conditions. Thus, a number of far reaching decisions needs to be made for the basic design of experimental set-up.

A fundamental decision for experimental investigations is the choice of the working medium. In many experimental studies, for instance in riblet research, liquids like commonly water or oil (Bechert *et al.*, 1997) instead of air are used as working fluid. Larger physical dimensions of the usually very small flow control devices exhibit advantages in manufacturing and larger friction forces are easier to be measured accurately in similar flow scenarios. This principle led to excellent accuracy in the case of the oil tunnel in Berlin (Bechert *et al.*, 1992).

The choice of the fluid influences the kind of drag reduction principles which can be investigated. Some of the drag reduction techniques are bound to liquid media such as long-chain polymers injection, which is not applicable in an air flow. In the same way the appliance of other flow control methods is limited to gaseous media. The aforementioned plasma actuator, for instance, ionizes gas molecules to produce a near-wall body force (Moreau, 2007) and is thus bound to the gaseous state. Also dielectric elastomer actuators (DEA) (Gouder (2011), Gatti (2014b)), tested in the context of the present work, cannot sustain the fluiddynamic and body forces in water flow, where the electric isolation is also problematic. Hence, certain flow control methods being of interest for the present investigation cannot be operated in liquids. Furthermore, air flow offers substantial advantages if the major problems of miniaturization and measurement accuracy can be resolved. In particular, the handling of a wind tunnel is simpler and less extensive maintenance is required in comparison to an oil or water tunnel facility.

In summary, there is valid reason to run experiments in air flow. However, the high accuracy determination of extremely small difference in the wall shear stress represents the challenging task of the present work.

Further, the type of investigated flow has a significant influence on the design of the facility and even on the appropriate measurement technique. There are three classical cases for experimental investigations of wall bounded flows: constant pressure boundary layer, pipe flow and channel flow (Schultz & Flack, 2013). While the boundary layer flow is characterized by a continuous development in the streamwise direction, both inner flows reach a fully developed state after a certain running length. This simplifies the reproduction of a flow in different experiments and facilities and hence their comparability. Constant pressure boundary layer at very controlled conditions as required in aerodynamic experimental investigations are more difficult to reproduce, which results in a limited measurement accuracy (Marusic *et al.*, 2010). Another crucial aspect is the need for detecting variations in the wall friction smaller than one percent. Oil film interferometry, which represents the most accurate technique available in zero pressure gradient boundary flow is generally specified with ($\pm 1 - 2\%$) (Marusic *et al.*, 2010). In addition the application of the oil film interferometry is constricted to smooth surfaces and thus not able to comply with the spectrum of interest for the research of the present work, which includes modifications of surface morphology.

However, studies in fully developed pipe and channel flows offer a precise determination of wall friction via pressure drop measurements along the length of a pipe or channel and the advantageous situation of extensive experimental (e.g. Dean (1978); Durst *et al.* (1998); Huebscher (1947); Monty (2005); Nikuradse (1932)) as well as numerical studies (Eggels *et al.* (1994); Jiménez *et al.* (2010); Kim *et al.* (1987); Moin & Mahesh (1998))- for the laminar case even analytical results (cf. Chapter 2) - that can be found in the literature (Schlichting, 1959; Spurk & Aksel, 2006).

The choice of a channel flow experiment in the present study enables comparison with numerical investigations of turbulent flow control. In a plenty of Direct Numeric Simulations (DNS) the convenient boundary conditions of such a quasi two-dimensional flow are chosen to economize computational effort. While the DNS provides a deep insight into details of the flow structures, experimental investigations can easily run extensive parameter studies. In ideal case numerical and experimental investigations work hand in hand to combine benefits and potentials of both approaches. From the experimental side a simple, but practical reason leads to prefer working in a channel: most techniques for flow control are complicated to apply on a curved surface of a pipe.

The 2D-channel flow is itself of minor interest for direct use in engineering application, but it serves as the key interface between the numerical and analytical research and the experimental studies. Additionally it exhibits favorable boundary conditions for all three disciplines of the research. Numerous experimental studies of riblets have proved that results from a channel flow (Bechert *et al.*, 1997; Grüneberger & Hage, 2011; Walsh, 1982) can be transferred to a boundary layer flow for turbulent conditions (Viswanath, 2002).

In summary, pressure drop measurements in channel flow allow high accuracy determination of skin-friction, containing precise information about changes in the wall friction and enabling to directly compare different techniques. These fundamental decisions on the experimental concept prescribe the approach taken in the present study.

In the first step an experimental facility is designed and built upon this basic concept, while in the second step it needs to be thoroughly verified, that this experimental set-up holds the expectations in measurement accuracy. Subsequently, application of selected active and passive techniques for skin friction reduction is to be demonstrated. Hence, the major goals of the present work can be formulated as follows:

Part I : Develop a platform to study and compare different techniques to reduce skin friction drag under the constraints of:

- accurate determination of skin friction based on the concept of pressure drop measurement in a channel flow
- applicability of active and passive control techniques
- ability to operate in laminar, transitional and turbulent flow conditions, while focusing on the latter
- flow medium air
- wide range of Reynolds number Re.

Part II : Demonstrate the capabilities of the set-up in comparison to literature data for a hydraulically smooth surface.

Part III: Show the applicability of active and passive techniques and enable new insight into previously investigated skin friction reducing approaches:

- riblets
- micro grooves
- spanwise wall oscillation

The riblet study has two goals. On the one hand the agreement to literature data can demonstrate the reliability of data generated in the facility of the present study. On the other hand the choice of air as working fluid allows to shed light on spatial transients, thanks to the longer running length compared to what is usually achieved in oil channels.

The few experimental as well as numerical studies on the topic of micro grooves leave room for various follow-up questions such as: Which is the optimal geometry? How does the optimal structure size scale with *Re*? Is it possible to extend the effective Reynolds number range? Additionally the micro grooves have been tested in transitional flows (Jovanović *et al.*, 2011), where further examination is required.

The key question regarding wall oscillations is how to efficiently implement the wall movement. The present facility turned out to be a very convenient test rig for the novel approach of dielectric electroactive polymers, that have been mechanically characterized and numerically studied by Gatti *et al.* (2014). In comparison to numerical simulations, the small surface area covered by the available actuators leads to very small differences in the skin friction. For the experimental set-up the challenge is to obtain a measurement accuracy which is high enough to well resolve the possibly very small variations of the skin friction drag.

The structure of this thesis is organized as follows:

In the first part the concept of the set-up will be developed from a theoretical standpoint. Subsequently, the implemented experimental set-up will be verified by data comparison with literature references. Therefore Chapter 2 introduces the fundamentals of a channel flow from analytical as well as from experimental side and adds numerical input where helpful. Additionally the essential properties of air are introduced. The following Chapter 3 argues the construction of the experimental set-up divided into the wind tunnel and the channel test section. The description of the set-up is completed with the introduction of the measurement instrumentation. In Chapter 4 a very thorough discussion of uncertainty estimation and corresponding error propagation is presented, which is of utmost importance for such high accuracy measurements. The measurement results of the channel with smooth walls are compared to literature data for fully turbulent and laminar-transitional flow in Chapter 5. Chapters 6, 7 and 8 discuss the concepts, the application to the channel experiment and present the measurement results of three different approaches for turbulent drag reduction: by passive means riblets (Chap. 6) as well as micro grooves (Chap. 7) and active represented by oscillating walls, implemented with novel type of dielectric electroactive polymer actuators (Chap. 8). The investigations are completed with a study of micro grooves in the transitional regime.

To the reader's convenience, the experimental studies for each flow control technique are written as self-contained reports.

2 Fundamentals of air channel flows

This Chapter introduces the fundamentals of channel flows from an experimental point of view. The Chapter starts with the definition of an ideal channel flow. On this basis analytical considerations of the channel flow will be made for turbulent as well as laminar flow conditions. Limitations of this approach due to the idealized boundary conditions are mentioned and the measurement principle for the determination of the wall shear stress is deduced. Additionally, the relevant properties of the fluid are discussed. The consideration is completed by the presentation of experimental and some numerical results characteristically observed in channel flows.

2.1 Analytical consideration of a channel flow

For analytical considerations a channel flow is defined as a steady flow in between two parallel plates of a distance H. The hypothesis is made that the plates are infinitely in length and width. In consequence the flow can be considered two-dimensional due to the infinite width and independent from the streamwise position due to the infinite length. Thus we can formulate following assumptions:

- $\Rightarrow$ the two-dimensionality of the flow field
- $\Rightarrow$ a fully developed flow state
- $\Rightarrow$ bounding walls are perfectly smooth

With regard to the experimental studies in air two more simplifications concerning the properties of the fluid are made:

⇒ for the low subsonic wind speeds of the present study the flow is treated incompressible. Furthermore even the density is assumed to be constant.

⇒ additionally air is assumed to be an ideal Newtonian fluid

While the definition of a Newtonian fluid, where the shear stress τ is proportional to the velocity gradient, can be directly confirmed to be appropriate for air (Spurk & Aksel, 2006), the condition for an incompressible flow - in particular the statement of constant density - requires a further examination. The condition for incompressibility of the flow can be deduced from the continuity equation

$$\frac{\partial \rho}{\partial t} + \frac{\partial}{\partial x_i}(\rho U_i) = 0. \tag{2.1}$$

Equation 2.1 can be rewritten to:

$$\frac{D\rho}{Dt} + \rho \frac{\partial U_i}{\partial x_i} = 0. \tag{2.2}$$

Hence, it needs to be shown that the simplification

$$\frac{D\rho}{Dt} = \frac{\partial \rho}{\partial t} + U_i \frac{\partial \rho}{\partial x_i} = 0 \tag{2.3}$$

is justified.

For the laminar and the turbulent conditions investigated in the present study it can well be assumed that the density is independent of time. This in turn implies that the density also needs to be homogenous to fulfill equation (2.3). To quantify changes in the density arising from the acceleration of the fluid, the ratio between the total density ρ_t of the resting fluid to the density ρ of the flow is determined using stream filament theory (Spurk & Aksel, 2006).

$$\frac{\rho_t}{\rho} = \left(\frac{\gamma - 1}{2} Ma^2 + 1\right)^{\gamma/(\gamma-1)}, \gamma_{air} = 1.4 \tag{2.4}$$

with

$$Ma = \frac{U}{\sqrt{\gamma R_s T}}. \tag{2.5}$$

If the static flow quantities are altered due to motion of the fluid such, that the effect is not negligible for the determination of the air density, a Mach-number of $Ma > 0.2$ is practically necessary. This means that a maximum deviation from the density of resting air of 2% is tolerated (Pope & Harper (1966) define even 300miles/h). In the present investigation the Mach number is limited to $Ma < 0.045$. Compressibility effects stay below 0.1% even for the highest Re. Therefore, the flow can be treated incompressible in very good approximation without any influence of the quality of the results. This still does not mean that the applicability of the assumption of constant density is valid. For the very accurate measurements in the present study even smallest changes of the static pressure and temperature through the channel can influence the measurement results, which is further examined in Chapter 4.

In the following the flow through a channel is discussed based on analytical considerations for laminar and turbulent conditions. Later on, limitations on the applicability of the remaining three assumptions for the experiment will be introduced step by step.

2.1.1 Laminar channel flow

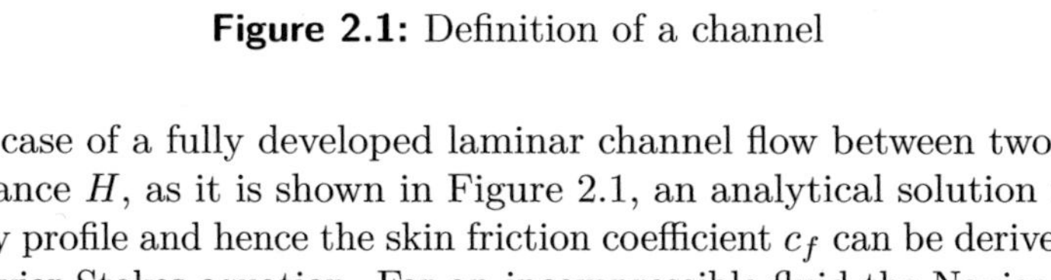

Figure 2.1: Definition of a channel

In the case of a fully developed laminar channel flow between two plates of distance H, as it is shown in Figure 2.1, an analytical solution for the velocity profile and hence the skin friction coefficient c_f can be derived from the Navier-Stokes equation. For an incompressible fluid the Navier-Stokes equations write:

$$\rho\left(\frac{\partial U_j}{\partial t}+U_i\frac{\partial U_j}{\partial x_i}\right)=-\frac{\partial p}{\partial x_j}+\eta\frac{\partial^2 U_j}{\partial x_i \partial x_i}+\rho g_j \tag{2.6}$$

In a steady air flow all derivatives in time vanish. The flow direction is in a horizontal plane and hydrostatic forces due to gravity are assumed to be negligible.

$$\frac{\partial U_j}{\partial t} = 0 \tag{2.7}$$

In a two-dimensional, incompressible flow the continuity equation writes:

$$\frac{\partial U_1}{\partial x_1} + \frac{\partial U_2}{\partial x_2} = 0 \tag{2.8}$$

Derivatives of U in x_1-direction vanish for a fully developed flow. With the boundary condition at the wall $U_2(x = \frac{H}{2}) = 0$ we obtain

$$U_2 = \text{const.} = 0. \tag{2.9}$$

The Navier-Stokes equations for the x_2-direction writes

$$0 = -\frac{\partial p}{\partial x_2} \Leftrightarrow p = p(x_1). \tag{2.10}$$

Thus the pressure is constant over a cross sectional plane at arbitrary position in the channel.
The x_1-component of the Navier-Stokes equations simplifies to:

$$0 = -\frac{\partial p}{\partial x_1} + \eta \frac{\partial^2 U_1}{\partial x_2 \partial x_2} \tag{2.11}$$

This differential equation is solved integrating two times in x_2. By doing so, two unknown constants C_1 and C_2 are obtained, whose determination succeeds concerning the constraints at the wall

$$U_{1(x_2=\frac{H}{2})} = U_{1(x_2=-\frac{H}{2})} = 0. \tag{2.12}$$

This enables to determine the flow velocity U_1 in streamwise direction in dependence on the channel height H, yielding to a parabolic shape of the velocity profile as shown in Figure 2.2.

$$U_1 = \frac{1}{2\eta}\left(-\frac{\mathrm{d}p}{\mathrm{d}x_1}\right)\left(\frac{H^2}{4} - x_2^2\right) \tag{2.13}$$

The velocity at the centerline U_c writes:

$$U_c = U_{1(x_2=0)} = \frac{1}{2\eta}\left(-\frac{\mathrm{d}p}{\mathrm{d}x_1}\right)\left(\frac{H^2}{4}\right). \tag{2.14}$$

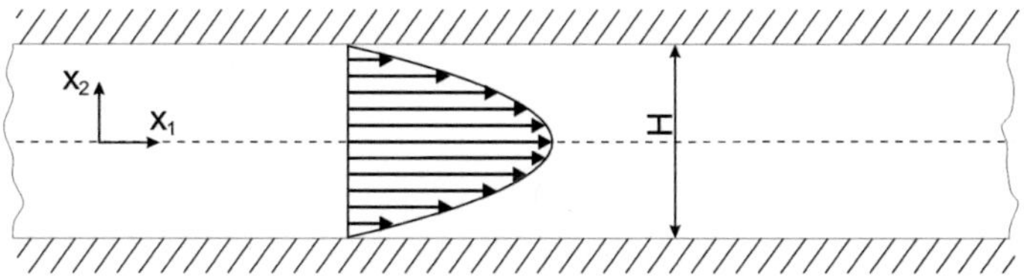

Figure 2.2: Parabolic velocity profile in a laminar channel flow

Now, the wall shear stress τ_w is introduced as

$$\tau_w = \eta \left.\frac{dU_1}{dx_2}\right|_{x_2=-\frac{H}{2}}. \tag{2.15}$$

By differentiating equation (2.14) with respect to x_2 and evaluating at the wall $x_2 = -\frac{H}{2}$ we derive

$$\left.\frac{dU_1}{dx_2}\right|_{x_2=-\frac{H}{2}} = \frac{1}{\eta}\frac{\mathrm{d}p}{\mathrm{d}x_1}\frac{H}{2} \tag{2.16}$$

and find the relation between the pressure drop and the wall shear stress

$$\tau_w = -\frac{H}{2}\frac{\mathrm{d}p}{\mathrm{d}x_1}. \tag{2.17}$$

Inserting this result in 2.14 allows to calculate the mean velocity U_b (b=bulk)

$$U_b = \frac{2}{H}\int_0^{\frac{H}{2}} U_1 \mathrm{d}x_2 = \frac{2}{H}\frac{1}{2\eta}\left(\frac{2\tau_w}{H}\right)\int_0^{\frac{H}{2}}\left(\frac{H^2}{4} - x_2^2\right)\mathrm{d}x_2$$
$$= 2H\frac{\tau_w}{\eta}\frac{1}{12} = \frac{2}{3}U_c \tag{2.18}$$

The skin-friction coefficient c_f is introduced by normalizing the wall shear stress with the kinetic energy of the bulk flow.

$$c_f = \frac{\tau_w}{\frac{\rho}{2}U_b^2} = \frac{\tau_w}{\frac{\rho}{2}(2H\frac{\tau_w}{\eta}\frac{1}{12})U_b} = \frac{12\eta}{\rho U_b H} = \frac{12\nu}{U_b H} \tag{2.19}$$

In the above equation the analytical results for τ_w (Eq. (2.17)) and U_b (Eq. (2.18)) in laminar channel flow have been used.

If we choose the channel height H as characteristic length to define the Reynolds number,

$$Re = \frac{U_b\, H}{\nu} \tag{2.20}$$

we get the simple expression for c_f

$$c_f = \frac{12}{Re_b}. \tag{2.21}$$

In fully developed channel flow the differential $\mathrm{d}p/\mathrm{d}x_1$ is constant and thus equal to the quotient of Δp and a length Δx_1. Considering the pressure difference Δp over a run length l, we can rearrange equation (2.17) and insert the expression we obtained for c_f in Eq. (2.21)

$$\Delta p = 2\tau_w\frac{l}{H} = \frac{12}{Re}\frac{l}{H}\rho U_b^2. \tag{2.22}$$

With this analytical expression for the pressure difference over a defined running length l, the resulting wall friction can be calculated for Reynolds numbers corresponding to the laminar flow state.

2.1.2 Turbulent channel flow

In turbulent flow we can assume the flow to be statistically steady and the fluctuations in all directions superposed to the mean quantities. Thus it is convenient to split the velocity in two components: A time-averaged mean value $\bar{U}$ and a fluctuation u.

Using this decomposition in equation (2.7) and (2.6) and time-averaging of the term yields the Reynolds-averaged equations (Pope, 2000) for continuity

$$\frac{\partial \overline{U_i}}{\partial x_i} = 0 \tag{2.23}$$

and momentum

$$\frac{\partial \overline{U_j}}{\partial t} + \overline{U_i}\frac{\partial \overline{U_j}}{\partial x_i} = -\frac{1}{\rho}\frac{\partial \overline{p}}{\partial x_j} + \nu\frac{\partial^2 \overline{U_j}}{\partial x_i \partial x_i} - \frac{\partial \overline{u_i u_j}}{\partial x_i} \tag{2.24}$$

in incompressible flow. Here, the last term describes the Reynolds stresses caused by the turbulent fluctuations. Now we follow the same procedure as used for the laminar flow and write the continuity equation for the mean quantities.

$$\frac{\partial \overline{U}_1}{\partial x_1} + \frac{\partial \overline{U}_2}{\partial x_2} = 0 \tag{2.25}$$

For a fully developed flow $\frac{\partial \overline{U_1}}{\partial x_1}$=0 we obtain

$$\frac{\partial \overline{U}_2}{\partial x_2} = 0 \Rightarrow \overline{U}_2 = \text{const} = 0. \tag{2.26}$$

With this result for $\overline{U}_2$ for wall-normal direction and the assumption of a statistically stationary and fully developed flow the second component simplifies to

$$0 = -\frac{1}{\rho}\frac{\partial \overline{p}}{\partial x_2} - \frac{\partial \overline{u_2 u_2}}{\partial x_2}. \tag{2.27}$$

Contrary to laminar flow we do not obtain a constant pressure over a cross section due to the additional stress term. We integrate 2.27 and determine the resulting constant defining the mean pressure at the top wall $p_w = p(x_1, \frac{H}{2})$ and insert the boundary condition at the wall $\overline{u_2^2}_{(x_2=\frac{H}{2})} = 0$:

$$\overline{u_2 u_2} + \frac{\overline{p}}{\rho} = \frac{p_w(x_1)}{\rho} \tag{2.28}$$

Differentiating in x_1-direction yields, that the mean axial pressure gradient is uniform across the flow.

$$\frac{\partial \overline{p}}{\partial x_1} = \frac{\partial p_w}{\partial x_1} \tag{2.29}$$

Now we write the x_1-component of the Navier-Stokes equations with the assumptions mentioned before:

$$0 = -\frac{1}{\rho}\frac{\partial \overline{p}}{\partial x_1} + \nu \frac{\partial^2 \overline{U_1}}{\partial x_2 \partial x_2} - \frac{\partial \overline{u_1 u_2}}{\partial x_2} \tag{2.30}$$

In contrast to the laminar case the problem is not analytically solvable. If we follow Pope (2000) and rewrite the equation using the expression for the total stress

$$\tau = \rho\nu \frac{\mathrm{d}\overline{U_1}}{\mathrm{d}x_2} - \rho(\overline{u_1 u_2}) \tag{2.31}$$

we obtain

$$\frac{\partial \tau}{\partial x_2} = \frac{\partial p_w}{\partial x_1}. \tag{2.32}$$

and find the gradient of normal stress balanced with the gradient of the total stress in x_2-direction.

The integration of equation (2.30) over half of the channel height $H/2$ with the total shear stress being zero for $x_2 = H/2$ yields:

$$
\begin{aligned}
0 &= -\int_0^{\frac{H}{2}} \frac{1}{\rho}\frac{\partial \overline{p}}{\partial x_1}\mathrm{d}x_2 + \int_0^{\frac{H}{2}} \nu \frac{\partial^2 \overline{U_1}}{\partial x_2 \partial x_2}\mathrm{d}x_2 - \int_0^{\frac{H}{2}} \frac{\partial \overline{u_1 u_2}}{\partial x_2}\mathrm{d}x_2 \\
&= -\frac{1}{\rho}\frac{\partial \overline{p}}{\partial x_1}\frac{H}{2} - \nu \left(\frac{\partial \overline{U_1}}{\partial x_2}\right)_0 .
\end{aligned}
\tag{2.33}
$$

As in the laminar case the pressure gradient $\partial \overline{p}/\partial x_1$ is in balance with the wall shear stress τ_w

$$
-\frac{\partial \overline{p}}{\partial x_1}\frac{H}{2} = \eta \left(\frac{\partial \overline{U_1}}{\partial x_2}\right)_0 = \tau_w . \tag{2.34}
$$

In summary it was shown that for laminar flow an analytical solution can be found for the velocity profile $U_1(x_2)$ and thus also for the wall shear stress τ_w. This allows the calculation the pressure difference due to friction losses over a distinct running length for a stationary, fully developed flow. In the case of a turbulent flow state an analytic solution exist neither for the velocity profile nor for the wall shear stress an analytic solution exists. With regard to the later presented measurement it is important to note, that the pressure difference between two axial position in both cases is a measure for the wall shear stress. In laminar flow the pressure is constant over the cross sectional area and can be read on arbitrary position of the cross section (cf. Eq. 2.10). In turbulent flow it needs to be considered, that a pressure distribution over the cross section exists (Eq. 2.28) and hence the pressure difference needs to be read at corresponding positions of the cross section. The topic of a pressure distribution spanwise to the flow direction will be examined experimentally in Chapter 5.

We obtain the same result from stream filament theory considering the channel from Figure 2.3. A filament of the length l starts at the cross section Ⓘ and ends at Ⓘⓘ. The Bernoulli equation extended for viscous fluids writes as follows

$$
p_I + \frac{U_{b_I}^2}{2\rho} + \Delta p_l + \rho g x_{2_I} = p_{II} + \frac{U_{b_{II}}^2}{2\rho} + \rho g x_{2_{II}}, \tag{2.35}
$$

where Δp_l is the pressure loss. The continuity equation delivers $U_{b_I} = U_{b_{II}} = U_b$. The x_2-position is the same for start and end of the flow filament. We derive:

$$p_I - p_{II} = \Delta p_l \tag{2.36}$$

The pressure loss due to wall friction is the only source to decrease the static pressure in flow direction. Thus the pressure drop is a measure for the wall friction in the channel.

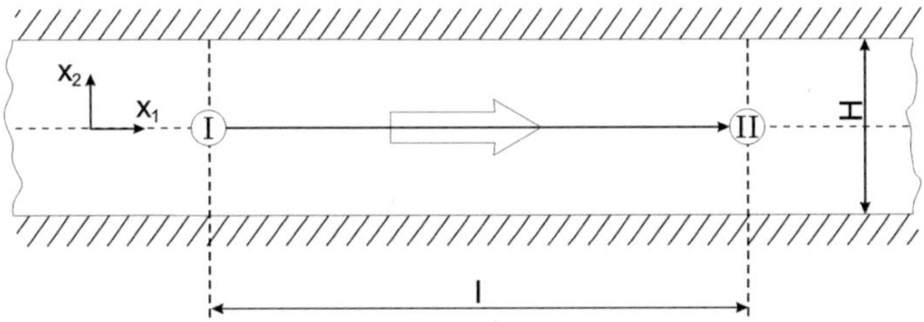

Figure 2.3: Consideration of pressure drop $\frac{\mathrm{d}p}{\mathrm{d}x_1}$ over a streamwise distance l based on flow filament theory

Law of the wall

The result for the turbulent channel does not offer as much physical insight as the complete laminar solution. However, some more information about the near wall region of a turbulent flow can be obtained by examining the near wall region via dimension analysis. Even in a turbulent flow the region in direct vicinity to the wall is dominated by viscous forces in a very thin layer. Fluctuations will be dampened due to the presence of the wall. Hence, the mechanism of momentum transport, similar to a laminar boundary layer, is dominated by viscous forces. With growing distance from the wall the inertial forces become dominate. To describe the functional relations of wall bounded turbulent flows the following dimensionless quantities are again derived from time-averaged mean values. The first quantity is the characteristic velocity for the near wall region named the friction velocity:

$$u_\tau = \sqrt{\frac{\tau_w}{\rho}}. \tag{2.37}$$

The viscous length scale is a measure for the smallest scales in the near wall region:

$$\delta_\nu = \nu\sqrt{\frac{\rho}{\tau_w}} = \frac{\nu}{u_\tau}. \tag{2.38}$$

These two characteristics quantities enable normalizing for velocity $\overline{U_1^+}$ and the wall normal coordinate x_2:

$$U_1^+ = \frac{\overline{U_1}}{u_\tau}, \tag{2.39}$$

$$x_2^+ = \frac{x_2}{\delta_\nu}. \tag{2.40}$$

Turbulent wall-bounded flows can be described by a universal law. As Figure 2.4 illustrates that the near wall flow is divided into to three layers. Close to the wall, where viscous forces are dominating, the velocity distribution follows a linear correlation $u^+ = x_2^+$. The region of validity of this linear law is named the inner layer or viscous layer for $x_2^+ \leq 5$ from the wall. In a distance of $5 < x_2^+ < 50$ from the wall a region of transition to the outer layer exists: the buffer layer. Farther away from the wall the outer or logarithmic layer begins. The velocity distribution follows the universal law (von Kármán, 1930):

$$u^+ = \frac{1}{\kappa}\ln x_2^+ + B. \tag{2.41}$$

This "Law of the wall" wall has been developed by von Kármán in a dimensional analysis. It has two constants, which were originally determined to $\kappa = 0.41$ and $B = 5.2$. However, the accuracy of determination is still discussed in the literature. Today's state of scientific knowledge indicates $\kappa = 0.38 - 0.39$ for the channel and boundary flow, while in a recent measurements for the Princeton super pipe $\kappa = 0.42$ was indicated (McKeon *et al.*, 1999). Marusic *et al.* (2010) debate whether the two values are indeed constants or slightly change with *Re*.

The linear and the logarithmic law deliver a description of the velocity profile near a smooth wall in a turbulent flow. This important result is utilized in the course of the work to examine the adaption of the wall-law to modifications of the surface condition.

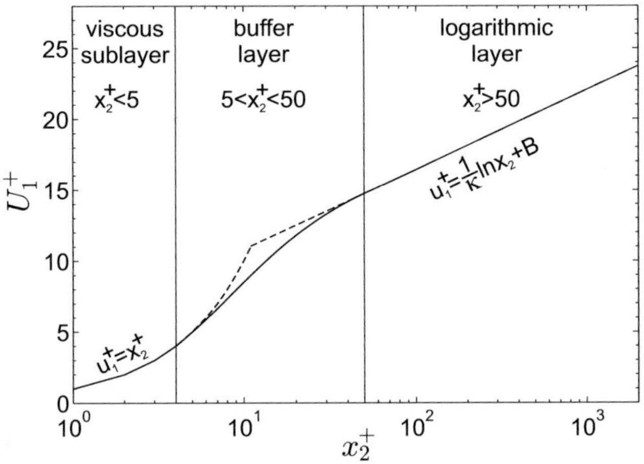

Figure 2.4: Velocity profile of a turbulent flow in the region near to the wall.

2.2 Examination of the initial assumptions for ideal channel flow

At the beginning of this Chapter the hypothesis of an ideal channel flow has been made to consider the analytic backgrounds. These assumptions are simplifications, which cannot be transferred to the experiment without examination. Some can be transferred to the experiment, while others can only be approximated. The following three assumptions require closer examination to justify or restrict their validity in the experiments.

- ⇒ the two-dimensionality of the flow
- ⇒ a fully developed flow state
- ⇒ bounding walls are perfectly smooth

Two of these demands on the channel flow turn out to be challenging for experimental investigations: The accomplishment of a full development as well as the two-dimensionality of the flow. While the full development can be achieved by an adequate length of the channel, two-dimensionality is limited to theoretical investigations. In theory it is possible to simply assume infinite plates, which is represented by periodic boundary conditions in numerical studies. But the physical realization of a channel of infinite dimension in spanwise direction is impossible. Hence for the experimental investigations the question of a sufficient ratio between channel height H and width W arises, which has been argued in numerous investigations in the past (Durst *et al.* (1998), Monty (2005), Marusic *et al.* (2010), Nagib *et al.* (2013), Schultz & Flack (2013)). The following section will present the state-of-the-art, focusing on the aspect ratio $AS = W/H$ and the development length L_E for turbulent as well as laminar flow conditions.

Further we still have a lack of knowledge about the turbulent channel flow since the analytic considerations do not provide a solution for the wall friction τ_w. Hence turbulent conditions require experimental and numerical data to complete the picture. The following section will discuss step by step these issues for laminar and turbulent flow conditions.

2.2.1 Concept of hydraulically smooth walls

One of the initial assumptions was the statement of perfectly smooth channel walls. In an experiment the question arises, which quality of the surface is convenient to approximate this criterion. This is a fundamental question as far as improper reference surface would lead to overestimation of the skin friction reduction on the test surface.

To quantify the influence of the surface property Nikuradse (1933) performed measurements in a pipe flow. The surface roughness has been varied using adhesive layers with sand grains of different sizes. For the smooth pipe Nikuradse found an excellent agreement with the law that was proposed by Blasius (1913). With Re based on the pipe diameter D the friction factor according to Darcy's definition writes:

$$\lambda = 4c_f = \frac{0.316}{Re^{0.25}}. \tag{2.42}$$

The comparison of the measurement with smooth and rough walls showed, that for a certain size of the sand grains, the friction in the pipe is increased, whereas it remains the same value as for the smooth pipe for smaller sand grain. Thus the concept of hydraulically smooth surface was formulated, where, if the sand grain size is expressed in terms of dimensionless units, Nikuradse's experiments indicated the flow to be unchanged by sand grain roughness height k for $k^+ = k/\delta_\nu < 3.5$. Over the years doubts towards this concept were growing for several reasons. The question arose whether random or distinct orientation of the surface morphology affects the influence on the flow. Also the nature of sand grain of roughness may not be the only parameter to characterize the surface. For example the structure of sand grain surface is to some degree affected by the binding material. Further missunderstanding arose due to Schlichting's interpretation (Schlichting, 1959) of Nikuradse's data, who stated the flow to be unaffected up to $k^+ = 5$. Additionally, the conclusion of Nikuradse was influenced by the measurement accuracy available at that time. In 2011 Grüneberger investigated roughness elements in form of small fins orientated spanwise to the flow direction (Grüneberger & Hage, 2011). He was able to accurately measure dimensions slightly below $h^+ = h/\delta_\nu = 1$ and influence on the flow was still found. The type of roughness in this experiments is untypical for industrially appearing imperfections, but a general limitation in the range of $k^+ = 3.5$ could be disproved. From these experiments the conclusion can been drawn, that the regime of hydraulically smoothness is found for $h^+ < 1$ - if such a regime exists at all. Bradshaw (2000) questions the existence of the hydraulically smooth regime in a passionate review. He concludes that there might be a boundary roughness below which the outcome of engineering applications is no longer affected by the surface morphology. However, from a scientific point of view, such a boundary roughness can probably not be found.

An estimate which has been shown to be appropriate for roughness type originating from industrial fabrication is the characterization of roughness based on the roughness height k to the pipe diameter D. Colebrook yielded a formulation for the Darcy friction factor in dependence of k/D (Spurk & Aksel, 2006):

$$\frac{1}{\sqrt{\lambda}} = 1.74 - 2\log\left(\frac{k}{0.5D} + \frac{18.7}{Re\sqrt{\lambda}}\right). \tag{2.43}$$

The result is found in the diagram by Moody & Princeton (1944) and Colebrook (1939) as shown in Figure 2.5 or the similar one by Nikuradse. The Moody-Colebrook diagram illustrates how roughness of a surface will enhance skin friction in a pipe flow. Within the turbulent regime roughness is more critical with increasing Re since the influence of a roughness of same size increases with Re and the criterion of a "hydraulically smooth surface" tightens. For $k^+ > 70$ the roughness significantly exceeds the region of the buffer layer. Then the friction factor stays constant with increasing Re (Spurk & Aksel, 2006), which defines a fully rough surface.

If we consider roughness in the layer to the law of the wall, roughness of size k will modify the velocity profile:

$$u^+ = \frac{1}{\kappa} \ln \frac{x_2^+}{k^+} + B = \frac{1}{\kappa} \ln x_2^+ - \frac{1}{\kappa} \ln k^+ + B. \tag{2.44}$$

The additional term $\frac{1}{\kappa} \ln k^+$ is a constant and the equation can be rewritten:

$$u^+ = \frac{1}{\kappa} \ln x_2^+ - \Delta B + B. \tag{2.45}$$

We obtain a reduction of the B modification corresponding to a shift towards the wall. Compared to the law of the wall for a hydraulically smooth surface the thickness of the viscous sublayer is reduced. For the fully rough surface the constant could be experimentally determined to $B = 8.5$:

$$u^+ = \frac{1}{\kappa} \ln \frac{x_2}{k} + 8.5. \tag{2.46}$$

In laminar flows roughness is of concern as it disturbs the flow triggering the transition process to turbulent flow. Since the flow velocities and the Re will be very small in the present experiments, the surface is considered to be smooth enough to keep laminar flow unaffected as long as the reference surface will comply to the stricter turbulent criteria. For the present investigation we will employ Grüneberger's finding (Grüneberger & Hage, 2011). Fulfilling the criterion $h^+ < 0.5$ for highest Re_b of the experimental study, the influence of roughness is assumed to be so small, that it does not noticeably influence the skin friction. Therefore the requirements on a "hydraulically smooth" surface can be fulfilled easily in the experimental investigation.

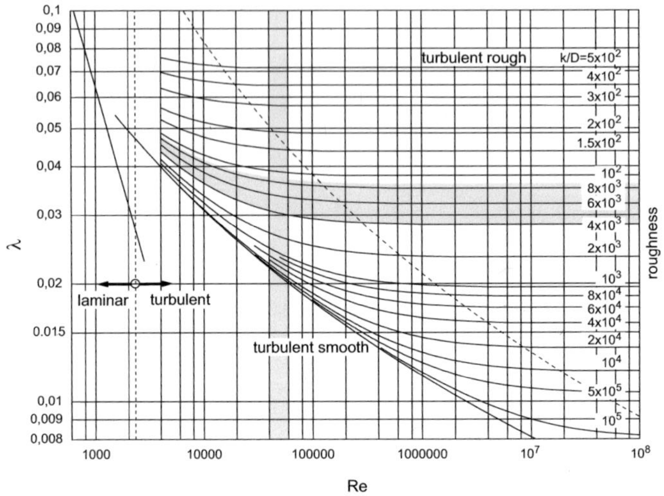

Figure 2.5: Friction factor λ (Darcy) in dependence of Re and the roughness parameter k/D for pipe flow following Colebrook (1939).

2.2.2 Flow development

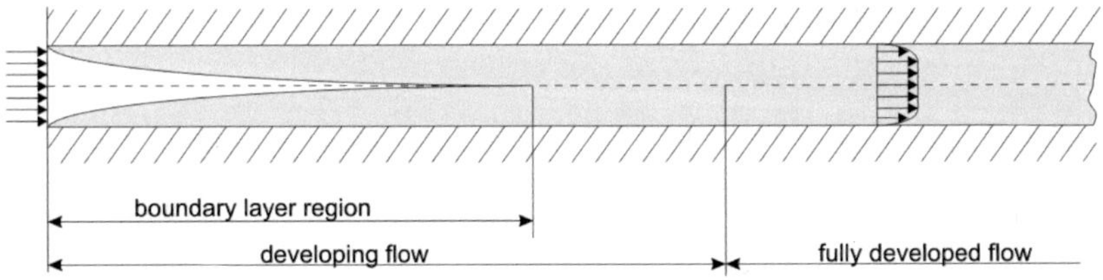

Figure 2.6: Laminar development of a 2D-channel-flow for low Re.

Any inner flow will need some distance to reach a state, where the velocity profile is independent of the streamwise position. A two-dimensional channel flow with smooth walls is shown in Figure 2.6. At the entrance the flow has a uniform velocity distribution. The velocity profile immediately reacts to the no slip condition at the channel walls. In streamwise direction boundary layers grow from both walls (illustrated in grey). Contrary to an open boundary layer flow, where in a distance from the wall the boundary profile merges into the free stream velocity, in a channel as well as in a pipe with increasing running length the growing boundaries layers meet at the centerline. A distinction between boundary layer and undistorted flow, as it is the case in a zero gradient boundary layer, is not possible. The complete cross-section is affected by the presence of the walls. In a fully developed state inertial forces do not play a role, since the flow is not accelerated or decelerated (Herwig, 2008). In developed flow the pressure gradient balances the friction forces (compare Eq. (2.17) and Eq. (2.1.2)).

We will first consider a flow which develops from the channel entrance without any disturbances. This will be denoted as a "natural" development which is illustrated in Figure 2.6. At low Reynolds numbers the flow develops laminar. Starting from a constant velocity distribution over the cross section at the entrance, the flow forms a parabolic profile at the fully developed state as shown in the analytic consideration in 2.1.1. In Schlichting (1934) the flow development in a laminar channel is examined theoretically. The boundary layer region is treated by the approximation of boundary layer growth on a flat plate according to Blasius for both walls. As the boundary layers have grown together the calculation follows an asymptotic development to the parabolic profile. The result is a fast growth of the required development length L_E with increasing Re for laminar flow. Schlichting's result was confirmed to be quite appropriate in several experimental studies, where L_E/H was found in a range of $0.055 - 0.68Re$ (an overview is given in Durst *et al.* (2005)). The study by Durst *et al.* (2005) found slightly longer development length and suggested the empirical relation:

$$\frac{L_E}{H} = (0.631^{1.6} + (0.0442Re_b)^{1.6})^{1/1.6} \tag{2.47}$$

which is illustrated in Figure 2.7.

With increasing Reynolds number inertial forces in the flow become dominant and the flow will become turbulent. At moderate Re the flow does not turn turbulent directly at the channel entrance as illustrated in Figure 2.8. It will take a certain distance until disturbances have grown large enough to initiate a transition to turbulent flow. The increase in lateral momentum transfer enhances the flow development. So the development length will be remarkably shortened compared to the laminar case. The laminar region at the entrance decreases with growing Re_b.

The laminar to turbulent transition process can be accelerated, if disturbances are induced through roughness of the surface or vibrations. Reynolds

(1883) observed the transition to occur between $Re = 2000$ and 13000 in a pipe, depending on the smoothness of the entry conditions. Usually $Re_{crit} = 2300$ is recommended for the onset of transition for a pipe and a flat channel. The critical Reynolds number Re_{crit} is not a fixed value, but a general value below that laminar flow will persist in engineering applications. A flow that can develop without any disturbance persists laminar up to significantly higher Re_{crit}. In this unstable state the flow promptly turns turbulent, if a for instance disturbed by an agitation.

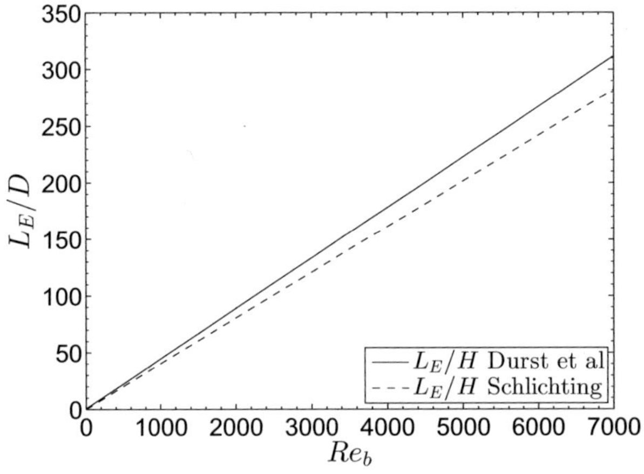

Figure 2.7: Development length of a laminar channel flow according to the theoretical consideration by Schlichting (1934) and the correlation derived from experimental data by Durst *et al.* (2005).

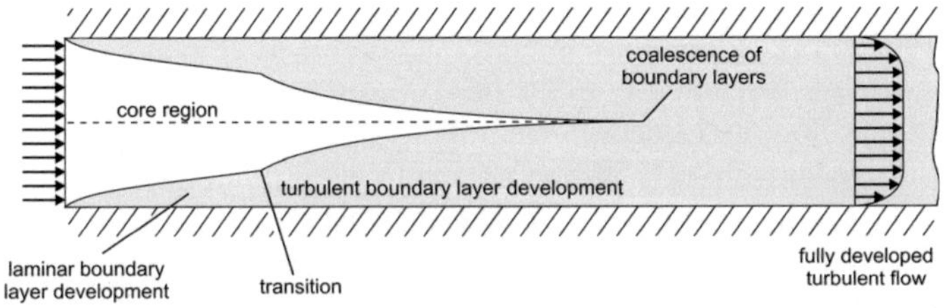

Figure 2.8: Natural development of a 2D-channel-flow.

Investigations under turbulent flow conditions are usually artificially disturbed to shorten the development length and to generate a turbulent development at low Reynolds numbers. The development length of such a tripped flow is controversially discussed, since the type and the size of the tripping device as well as the definition of a fully developed turbulent flow differ between the experimental studies. From an asymptotic approach in Herwig & Voigt (1995) a development length of $L_E = 25H$ is indicated to be sufficient in a channel. This result is in accordance with experiments from Byrne *et al.* (1970), while Durst recommends $60H$ (Durst *et al.*, 1998). For a turbulent flow commonly the development length is recommended to be independent of the Reynolds number ($L_E \neq f(Re)$).

The effectiveness of such tripping devices is limited by the Reynolds number range, beyond which the flow is totally dominated by viscosity. At very low Re Nishioka & Masahito (1984) found even large distortions to be promptly dampened and the flow to return to the laminar state. The minimum Reynolds number beyond that disturbances can sustain over a long distance in a channel was obtained about $Re = 1400$. This result is confirmed very accurately by further experimental studies (Davies & White (1928), Carlson *et al.* (1982)) and also by the analysis of secondary instabilities by Orszag & Patera (1983). The lowest Re to achieve a fully turbulent developed flow state is recommended by Durst *et al.* (1998) with a centerline Reynolds number of $Re_c \approx 3000$, corresponding to $Re_b \approx 3500$, which is in agreement with Nikuradse (1932). Nishioka & Masahito (1984) achieved fully turbulent conditions with a strong tripping for $Re_b = 2400$ in a very long channel ($L = 400H$).

One important aspect to allow to transfer measurement results from a channel flow for general application in turbulent flows, is the question which Reynolds number is required to achieve a flow state, where the gradient at the wall will follow the log-law distribution. Nishioka & Masahito (1984) found the flow to partly follow the universal law for $Re_b > 3000$. Durst *et al.* (1998) found a universal behavior in the near all region (linear law) even for small turbulent Re, but for $Re_b \leq 3000$ their data show considerable deviation for the region of the logarithmic law.

2.2.3 Approximation of two-dimensionality

As introduced at the beginning of this chapter, ideal two-dimensional channel flow is useful for the theoretical methods. If we speak about a channel flow in the experimental context, an approximation with the geometry of a flat duct is meant. A flat duct with the width W and the height H is characterized by the aspect ratio $AS = W/H$. The fundamental difference of a turbulent flow in a flat duct to the flow between infinite plates is illustrated in Figure 2.9. In a square duct secondary motions form which transport momentum to the corners. The result is a nearly constant distribution of the wall shear stress in turbulent flows (Spurk & Aksel, 2006). On the one hand these corner vortices balance the shear stress which is favorable in context of a two-dimensional skin friction distribution in spanwise direction. On the other hand these secondary motions significantly alter the flow field in the corners. For the flat duct the region influenced by the secondary motions is limited to region of the sidewalls. Furthermore, additional friction is generated at the side walls compared to the flow between infinite plates.

The question arises on the aspect ratios needed to neglect the influence of the flow in vicinity to the sidewalls. In the literature differing estimates are given depending on the definition of what can be accepted as still two-dimensional. While Dean found $AS \geq 7$ satisfactory, Schultz & Flack (2013) use $AS = 8$, Marusic *et al.* (2010) states $AS \gg 10$, Nagib recommends $AS \geq 24$ (Nagib *et al.*, 2013) and (Durst *et al.*, 1998) as well as Monty (2005) find $AS = 12$ adequate. The question of a sufficient approximation of a two-dimensional channel flow is closely connected to the question of a sufficient development length. Experimental data of high aspect ratio together with large channel length are rare.

The contribution of numerical simulation to the topic on high aspect ratio ducts is insufficient. The DNS simulation exhibiting the highest aspect ratio $AS = 7$ known to the present author is the recent studies by Vinuesa (Vinuesa, 2013; Vinuesa *et al.*, 2014). From the presented visualization of crossflow streamlines and magnitude it can be deduced, that an aspect ratio of seven is definitely too low for the small $Re_b \approx 2800$. As will be shown in Chapter 5 a real channel flow does not fulfill the criteria of a fully turbulent flow at such low Re_b. With increasing Reynolds number the momentum exchange is also increasing and the flow field is rather approximating a two-dimensional state.

For practical reason a ratio of $W/H = 12$ is chosen, which seems to be useful in the present case. To get a profound information about the degree of deviation from a two-dimensional state, the spanwise distribution of the skin-friction has been measured and is discussed in Chapter 5.

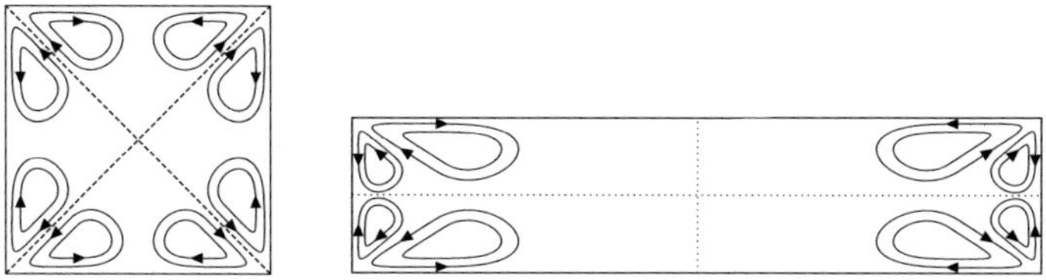

Figure 2.9: Schematic illustration of secondary motions in a square (left) and a rectangular duct (right) according to (Hoagland, 1960).

Influence of non-ideal 2D-flow on the skin friction coefficient

A fundamental study of experimental channel flow is the one from Dean (1978), who collected a lot of data from different investigators. The available data had been generated in facilities of differing characteristics. Finally 27 sources dating from 1928 to 1976 were taken into account. Partly data needed to be recalculated to illustrate all data in one diagram showing friction coefficient c_f over Re_b. Based on this data collection Dean developed his famous fit for the friction factor in an experimental channel flow:

$$c_f = 0.073\, Re^{(-0.25)} \tag{2.48}$$

The data collected by Dean show large scatter. Besides varying measurement accuracies of the different sources, the data is influenced by the variation in the aspect ratio, reaching from $AS = 7$ to 169. Additionally there is another influence that affects some of the data in Dean's data base: the length for flow development. For example Laufer (1951) and Skinner (1951) used a short channel with a length of $55H$ and let the flow develop only $20H$, which seems not appropriate, if we recall the values given by Herwig & Voigt (1995) and Durst *et al.* (1998). Yet, most sources used channels exhibiting $100 < L < 300H$, for instance (Huebscher, 1947), some even $L > 300$ for example (Hartnett *et al.*, 1962). Dean himself was aware of these facts and states both. Marusic *et al.* (2010) remark that the development length is one of the critical criteria leading to disagreement in available data and that there is still no common agreement concerning the sufficient value. Nontheless, Dean‘s correlation represents a widely accepted approach which agrees very well with wide range of experimental studies.

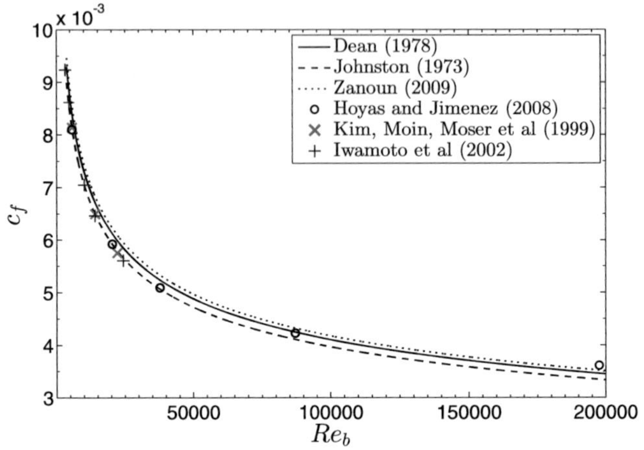

Figure 2.10: Friction coefficient c_f: Correlations from experimental data (Dean, 1978; Johnston, 1973; Zanoun *et al.*, 2009) in comparison to DNS results by Hoyas & Jiménez (2008); Iwamoto *et al.* (2002); Moser *et al.* (1999).

Recently a refinement of Dean's correlation was intended by Zanoun *et al.* (2009). Zanoun's fit (Eq. 2.49) needs to be discussed here since it is based on measurement data from a $AS = 12$ test section similar to present channel.

$$c_{f_Z} = 0.0743\, Re^{(-0.25)} \tag{2.49}$$

Indeed the new correlation showed better agreement to the data of Zanoun *et al.* (2009) as well as to the data of Monty (2005) than Dean's law. The cited sources are mainly interested in the investigation of universal flow quantities regarding high-*Re* flow. Thus the developed empirical formula for the friction coefficient focuses on $Re_b \gg 20.000$, while the present work will concentrate on comparably low-*Re* flows which allows a comparison to numerical simulations. For the range $Re_b < 80000$ still Dean yields the better fit to the mentioned experimental data. Yet, it should be realized, that experimental facilities can never provide a perfect two-dimensional flow. Evidence for the magnitude of the deviation to an ideal two-dimensional flow can be obtained by considering DNS data. Kim *et al.* (1987) results show 3% lower c_f-values in comparison to Dean's correlation in the low-Re range. The deviation becomes smaller with increasing Re. Figure 2.10 compares the correlation from Dean, the older one from Johnston (1973) $c_{f_J} = 0.0706\ Re^{(-0.25)}$ and the recent fit by Zanoun. The data points indicate the results originating from DNS. Note that all correlations use the same functional relation $c_f = k\ Re^{(-0.25)}$ and only modify the constant k. It can be conjectured that it is impossible to match the experimental data for the whole range of Re on that basis. The resulting value of the constant is a function of the Reynolds number.

Influence of secondary motions on flow development

As shown in Figure 2.9 secondary motions influence the turbulent flow in a flat duct. Studying laminar flow through a flat duct Arbeiter (2009) showed a successive transport of fluid from the side wall region to the center of the duct. In personal communication Arbeiter mentioned that this mechanism was still not finished over quite large run length (L_E=80). This may explain the slight difference in L_E reported between Schlichting's theoretical and experimental studies (cf. Chapter 2.2.2). A special issue is the effect on laminar-turbulent transition in the duct. In an ideal two-dimensional flow the transition process is initiated by small disturbances that are developing with the run length as it is the case in a pipe flow experiment. However, in a flat duct disturbances unavoidable appear in the form of the corner vortices, which are growing stronger with increasing Re. Thus secondary motions can induce transition starting from the "inflectionally-unstable corner boundary layers" (Dean, 1978), which is distinctively visualized in Arbeiter *et al.* (2008). Arbeiter reports a drastic increase of turbulence within the streak regions at the side walls.

Since no particular information for the development of the vortex dominated region at the sidewalls for a flat duct was found in the literature, an estimate of the influence of the sidewalls on the transition process is drawn from the flat plate experiments by King & Breuer (2001, 2002). Their wind tunnel exhibits a conventional layout of H = 2ft and W = 3ft with a flat plate located horizontally in the center (Fig. 2.11). King & Breuer (2001) illustrate a contamination with an angle 12° on the plate originating from the side walls. If we consider the cross section being divided by the flat plate, the obtained sidewall contamination can be transferred to each of the two flat ducts with an $AS = 3$. The angle of 12° is in agreement with the photograph shown in Arbeiter (2009) from a channel exhibiting $AS = 10$. It can be deduced that for moderate aspect ratios the sidewall contamination is influencing the transition process.

As a consequence a hydraulically smooth duct will have a certain Re where transition is occurring (Dean, 1978). This situation is fundamentally different to the pipe, where the condition of the inlet is the main parameter influencing the transition process. If not disturbed, a pipe flow can remain laminar up the very large Re.

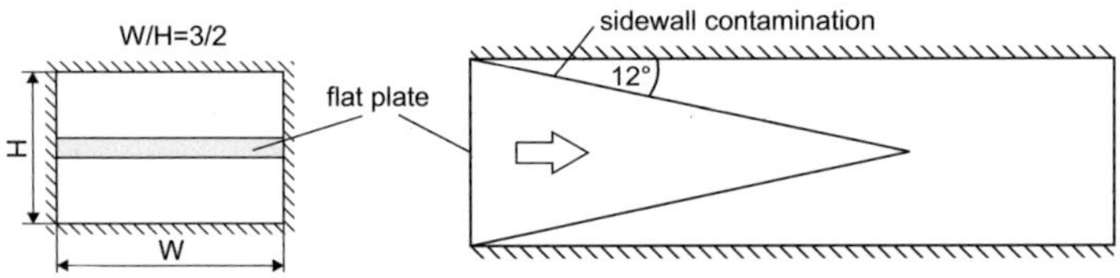

Figure 2.11: Schematic sketch of side wall effect in the wind tunnel experiment of King & Breuer (2001), including cross sectional and top view.

2.3 Properties of air

In terms of aerodynamic considerations certain characteristics of the gas mixture air – mainly consisting of nitrogen and oxygen – are from relevance. For the present study the density ρ and the dynamic viscosity η are of major interest.

In the first instance we regard the properties of a fluid employing the ideal gas law (Eq. 2.50). With the static pressure p and ambient temperature T as well as the specific gas constant R_s, the density can be determined. Using the standard conditions from table 2.1 defined in White (1974), corresponding to German engineer standard DIN2533 (Czichos & Hennecke, 2004) we yield a density of ρ =1.225kg/m^3. The right graph in Figure 2.12 depicts the behavior of the air density at norm temperature in range of the meteorological fluctuation of the atmospheric pressure, while the left one shows the equivalent for temperature variation at a standard pressure of 1013.25hPa

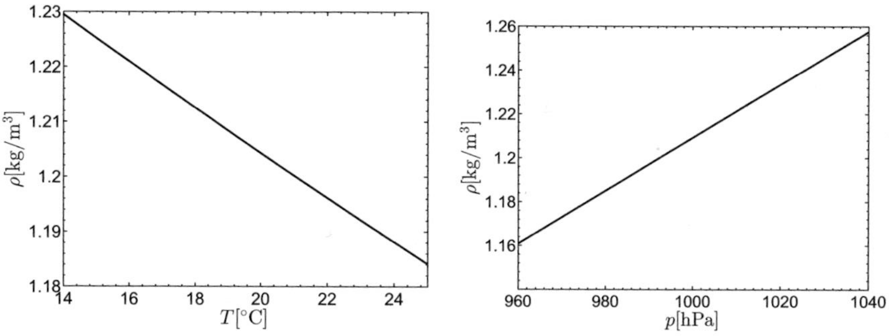

Figure 2.12: Air density within the temperature range of the laboratory (left) and in dependence of typical variations of the atmospheric pressure (right).

$$p\,V = n\,R\,T = R_s\,T \qquad \Leftrightarrow \qquad \rho = \frac{p}{R_s T} \tag{2.50}$$

Additionally, the density is affected by the humidity in the air (Fig. 2.13). The air humidity can be taken into account by determining the gas constant of the air-vapor mixture R_f. Following the empirical formula (2.52) from Magnus, the saturation vapour pressure $p_{d,s}$ is calculated. The air humidity h is inserted in the common form as the percentage of vapor which the air can maximal contain. Further the specific gas constants for air R_l and vapor R_d are required

$$\rho = \frac{p}{R_f T} \text{ with } R_f = \frac{R_l}{1 - \left(h\frac{p_{d,s}}{p}\right)\left(1 - \frac{R_l}{R_d}\right)} \tag{2.51}$$

$$p_{d,s} = p_{d,s,0} \exp\left(\frac{17.5043 \cdot (T - 273.15)}{T - 31.95°K}\right). \tag{2.52}$$

Also the dynamic viscosity η is a function of p and T:

$$\eta = f(T, p_{bar}). \tag{2.53}$$

As depicted in the diagram 2.14(a), η is weakly dependent on the barometric pressure. For instance a pressure increase from standard pressure to 5bar

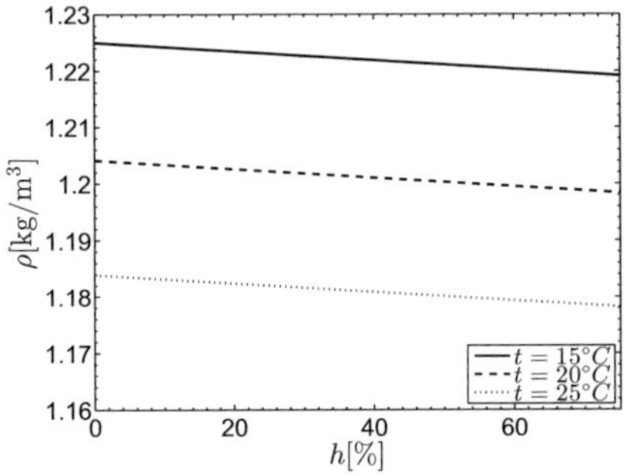

Figure 2.13: Dependence of air density on the humidity for different temperatures.

air at normal state (DIN ISO 2533)		
standard pressure	p_n	101325Pa
specific standard temperatur (DIN 1343)	T_n	228.15K=15° C
density of dry air	ρ_n	1.225 kg/m^3
specific gas constant of dry air	R_l	287.05287J/(kgK)
dynamic viscosity	η_n	17.894μPa·s
specific gas constant of vapor (VDI 2002)	R_d	461.526J/(kgK)
Sutherland constant (White 1974)	S	110.556

Table 2.1: Standard conditions of air according to DIN ISO 2533, Sutherland constant S and specific gas constant of water steam R_s

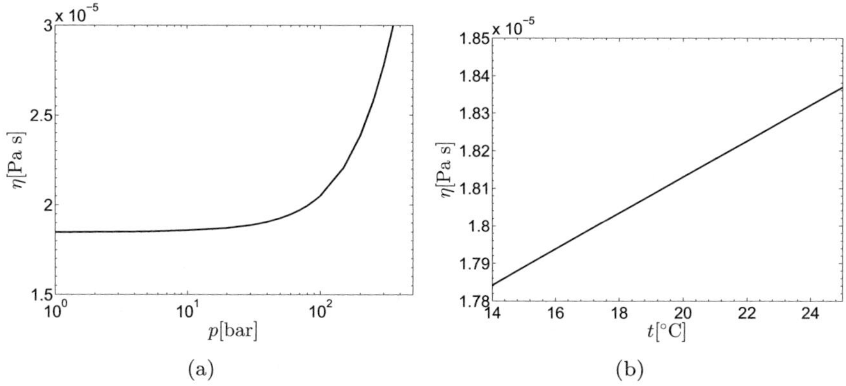

Figure 2.14: Air viscosity as a function of the pressure according to VDI (2002) (a) and in dependence of the temperature in the laboratory (b).

alters the viscosity by only 0.2%. Taking into account the small fluctuations of the atmospheric pressure of $\Delta p_{bar,max} = \pm 40$hpa the effect on η neglected. In consequence the viscosity solely depends on the temperature, so the variation can be determined according to Sutherland's formula (Eq. 2.54 from White (1974)) as shown in Figure 2.14(b).

$$\eta\left(T\right) = \eta_0 \frac{T_0 + S}{T + S} \left(\frac{T}{T_0}\right)^{\frac{3}{2}} \tag{2.54}$$

The Sutherland constant S and the reference temperature T_0 also correspond to the definition from White (1974) in table 2.1.

Depending on the literature source, differing information at various reference points are specified for standard parameters of air. If Sutherland's formula is used to convert to a concerted reference point, differences in the calculated air viscosity result. Values given in White (1974) differ from VDI (2002) about 0.5%. If we consider that the viscosity affects the Reynolds number proportional to $1/\eta$, this deviation is of the order of the errors in the flow rate measurement presented in Chapter 4. In the present study a potential systematic deviation would not affect the result of the percentage reduction of skin friction, but it shows that the reference source needs to be chosen carefully. The author cannot say which source is the most exact. The reference values of White (1974) are chosen to stay conform with other aerodynamic research.

3 Experimental setup

The following section presents the experimental setup, that has been designed and built within the project. The discussion will be divided in three main parts. The first part provides a detailed description of the wind tunnel, while the second part considers the test section. Part three introduces the measurement equipment and examines the interaction of the complete system focusing on the capabilities of the set-up in combination with the channel test section. It is followed by a detailed estimation of the measurement uncertainty in the subsequent chapter.

3.1 Wind tunnel

In the context of the present project special requirements on the wind tunnel are addressed, which are found in a configuration denoted "blower tunnel", where the fan is installed upstream the test section. The blower configuration is advantageous with respect to effective wind tunnel control and uniformity of the flow rate (Pope & Harper, 1966). In the particular case of the present project, the type of a wind tunnel in its specification as a modification of a ventilator test rig according to DIN 24163 (1985) is recommend due to requirements in highly accurate flow rate measurement and large pressure built-up. Facilities of this type are widely-used and often modified to extend the initial intended field of operation. For instance facilities of this type run at institutes FSM at KIT in Karlsruhe and LSTM at Friedrich-Alexander-Universität in Erlangen (Spudat, 1981). In contrast to the present set-up, these facilities were designed and built for comparably high flow rates. Both serve as basic instrument for scientific research for many decades. LSTM in Erlangen provided the opportunity to inspect their facility, which is intermittently used in very similar experiments with a comparable test section configuration as the one designed for the present study.

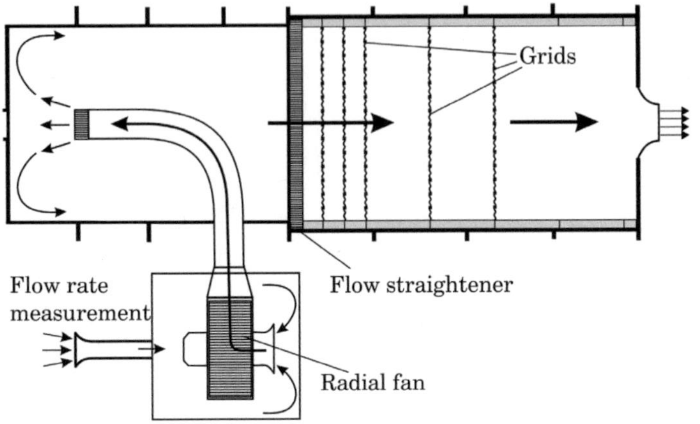

Figure 3.1: Functional structure and working principle of the blower tunnel presented on base of sectional top view on the facility. The air flow is symbolized by arrows.

Deeper insight into the topic of blower wind tunnels is given by Mehta & Bradshaw (1979). The authors supply a detailed analysis of the blower tunnel principle including a load of practical and useful information. Further attention to the topic is paid by Bradshaw & Pankhurst (1964), who provide some helpful drawings of existing wind tunnels, which are working on base of the blower principle.

A major aspect of the layout of the wind tunnel is the choice of the fan. While the present project requires small flow rates in the range of $40 - 400\text{m}^3/\text{h}$, the range of operation for future projects was planned to reach $30000-40000\text{m}^3/\text{h}$. Since a single fan providing such a wide operation range is not available on the market, the facility has been designed and constructed for high flow rates, but is equipped with a "small" blower. Nevertheless, a large operation range of $40 - 6000\text{m}^3/\text{h}$ is possible (all specifications are given for air at standard conditions referring to table 2.1).

Figure 3.1 illustrates the functional principle of the facility. An instrument for measuring the instantaneous flow rate is installed at the inlet of the tunnel. Air is sucked though a inlet nozzle (mapped) or an orifice meter depending on the flow rate. The flow is driven by a radial ventilator controlled by a frequency converter (FC). The air is led through a supply pipe into a large settling chamber. On the way through the settling chamber the air passes a flow straightener and several grids. Dependent on the requirement of the investigation open jet nozzles or test sections of variable cross section - in the present case a flat channel - can be attached to the outlet of the settling chamber. The system is controlled by a National Instruments (LabView) interface, which monitors the flow rate and transfers analog control signals to the frequency converter.

The following detailed discussion of the present wind tunnel is composed of four functional modules:

- radial ventilator
- settling chamber
- flow rate measurement
- wind tunnel control.

3.1.1 Radial ventilator

A radial ventilator with an electric power of 5.5kW (Ziehl-AbeggRG45-T-2DN.G5.2R) serves as the propulsion unit and is located in a separate housing at the suction side upstream of the settling chamber. Hence maintenance and repair are facilitated. The ventilator is operated via a Siemens Micromaster 440 frequency converter in a range of $4-50$Hz or $240-3000$rpm, respectively. The rotational frequency can be adjusted in steps of 0.02Hz corresponding to approximately 1rpm. The maximum available rotation range is highly depending on the pressure loss generated by the attached test section and the flow rate instrument. In combination with the channel with small cross section, used in the present study (cf. Chapter 3.2), the surge line is reached at approximately 25Hz for all flow rate instruments and the blower is limited to 15% of the maximum flow rate. The full range can be utilized if a free jet nozzle (pictured in Fig. 3.1) or test section of moderate cross section is attached. This is shown by Holderried (2011), who demonstrates the capability available for future projects. In general, the characteristic curve of the facility will modify in any combination of wind tunnel and test section. Therefore, the discussion will concentrate on the setup used for the drag reduction experiments. One can interpret the whole study as a determination of the characteristic line of the wind tunnel in combination with a variable resistor, represented by the test section in modified conditions.

3.1.2 Settling chamber

The settling chamber consists of 4mm thick cylindrical steel casing with ribbing for reinforcement configured for a system pressure of $\Delta p_{max} = 0.05$bar. The length is $L = 4800$mm and the inner diameter is constant ($d_i = 1760$mm). The inflow of the air is implemented via a supplying pipe ($D_R = 400mm$), which directs the air to the headwall of the chamber (compare Figure 3.1). The arrangement to blow the air towards the end wall is adopted from the facility in Erlangen (Spudat, 1981). It may not seem intuitive to blow contrary to the main flow direction into the chamber. Alternatively, the length required to expand a free jet, which is blowing upstream in the chamber, to a homogeneous velocity distribution over the diameter of the settling chamber needs to be considered. If the installation of auxiliary components such as grids or perforated plates to shorten this length would at least lead to less pressure losses in the settling chamber remains questionable.

In the downstream section of the settling chamber the diameter of the steel casing expands to gain additional space for a cribbing with wooden rings (glued laminated beech 30mm, $d_i = 1760$mm). The wooden rings carry 5 grids (d=2mm, blockage 56%, stainless steel) and fix the flow straightener (TUBUS honeycomb PC6.0 − 70 depth 100mm). By opening the end cover on the outlet side the rings can be removed or exchanged. This enables to vary the distance between the grids or to equip the rings with meshes of different width.

The functionality of the settling chamber needs to be considered for two different ranges of the flow rates. For small flow rates ($< 900\text{m}^3/\text{h}$) the mean velocity in the chamber remains smaller than $< 0.1\text{m/s}$. Thus, the chamber is practically a settled plenum. If the operating range of the facility is expanded to the maximum capabilities for flow rates of up to $40.000\text{m}^3/\text{h}$, the grids become effective.

Again the arrangement of the ventilator outside the settling chamber has important benefits. On its way through the settling chamber the flow does not pass the blower, but only the supplying pipe with a comparable small blockage. Additionally the positioning of the drive unit is favorable to decouple the vibrations and to some extent the noise generated by the ventilator. On the other hand the selected arrangement turns out to be disadvantageous due to the larger distance the flow has to cover from the inlet to the outlet of the facility. The resulting pressure losses, in particular at the low flow rates of the actual operation range, are insignificant.

The selected arrangement has been chosen to generate a homogeneous and symmetric velocity distribution at the outlet. Holderried (2011) verified the homogeneous distribution of velocity and low turbulence level in a free jet configuration for various outlets diameters and flow rates.

The accessibility of the settling chamber for repair, maintenance or cleaning of the inside is of fundamental importance for the everyday work. Therefore, the facility offers two possibilities to enter the settling chamber on each side: The access through the window at the bow or a removal of the complete 900mm-section at the inlet side. On the outlet side of the chamber one can enter through the standard flange with d_{out} = 800mm or, as mentioned before, open the full diameter by unfastening the front cover. The common uncomfortable and costly construction with a manway opening is avoided.

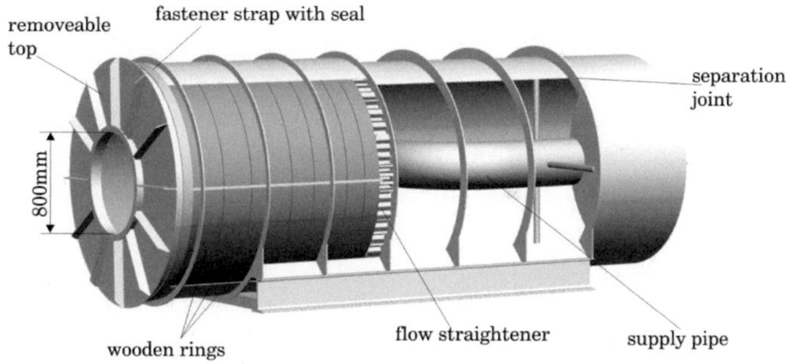

Figure 3.2: Illustration of the settling chamber including partly a sectional view to depict the inner workings.

3.1.3 Flow rate measurement

In general the working principle of the selected types of flow rate meters is based on the measurement of the differential pressure between the static pressure in the flow in a pipe and the total pressure at a stagnation point. The obtained differential pressure is proportional to the square of the fluid velocity. The flow rate $\dot{V}$ can be determined as follows (DIN24163-2):

$$\dot{V} = k\,A\sqrt{\frac{2\Delta p}{\rho}}. \tag{3.1}$$

The constant k is dependent in the expansion number e and a typical flow coefficient, which is a function of the Reynolds number and the characteristic geometry depending on of flow meter type. A is the cross section area of the instrument.

In the present study the wind tunnel can be equipped with two types of differential pressure based flow rate instruments. Depending on the flow rate inlet nozzles or orifice meters are chosen for the flow rate determination. Both types of flow rate meters are used in combination with two Setra D239 sensors with different measuring ranges and an accuracy of 0.07% of their full scale (FS). Since the pressure difference at the flow meters for both types of instruments to a large portion generates dissipative loss, it can have a considerable influence on the operation range of the facility. In order not to limit the capabilities of the facility by generating high pressure loss in the flow measurement, a maximum differential pressure of $\Delta p_{max} = 625\text{Pa}$ was selected. Due to the accuracy of the sensors, the lower limit of reading was restricted to an operation range of $\Delta p = 18 - 125\text{Pa}$ and $123 - 625\text{Pa}$.

Inlet nozzles

Inlet nozzles are widely used devices for flow metering complying with engineer standards DIN 24163 (1985) and DIN EN ISO 5167-2 (2003). High accuracy, small space requirement, simple manufacturing and comparably low pressure losses at high flow rates characterize these instruments. From the different types of inlet nozzles the construction following DIN24163-2 “Viertelkreismessdüse” was chosen (Figure 3.3). The version with four

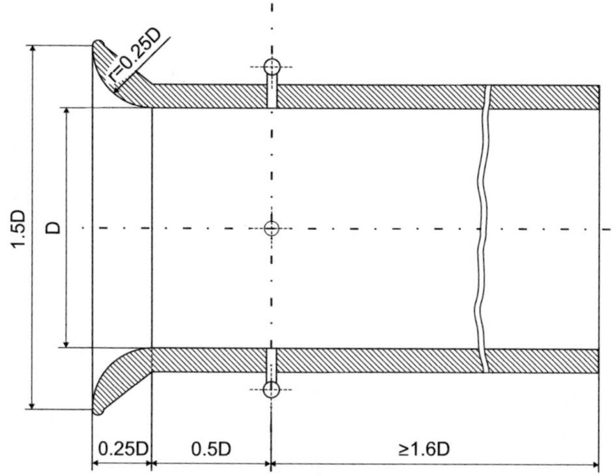

Figure 3.3: Cross sectional view of an inlet nozzle of the chosen type with quarter cycle inlet.

pressure taps of a diameter of 0.3mm placed equidistantly over the circumference was selected. Five nozzles of diameters of 50mm, 74mm, 112mm, 168mm and 262mm have been built at the workshops of the institutes SLA and PTW in Darmstadt. In combination with the Setra sensors an operation range of $\dot{V} = 200 - 5300\text{m}^3/\text{h}$ is covered.

The determination of the flow rate is based on a measurement of the static pressure difference between a certain position in the nozzles and the ambient pressure p_{bar}. As illustrated in Figure 3.3 the pressure tab is located at a distance of $0.5D$ from the inlet. At this position a minimum development of the flow can be assumed and the cross sectional velocity distribution is nearly constant. As far as a turbulent flow state is achieved the flow coefficient α is delivered by a calibration function according to DIN:

$$\alpha = 1 - 0,004\sqrt{\frac{10^6}{Re_D}} \quad \text{with} \quad Re_D = \frac{U_b\, D}{\nu}. \tag{3.2}$$

To obtain turbulent flow at the tap position, $Re_D \geq 100000$ is required. Thus for a certain nozzle diameter D a lower limit of the operation range results. In general a nozzle of smaller diameter enables a lower minimum flow rate. But smaller diameters result in larger velocities to reach the Reynolds number requirement. In consequence a very small nozzle would be a pure generator of pressure loss due to increased friction. Therefore, DIN 24163 (1985) formulates a practical limit of $D_{min} = 50\text{mm}$ corresponding to approximately $\dot{V} = 200\text{m}^3/\text{h}$. Since the specification for the present project was to measure flow rates down to $\dot{V} = 40\text{m}^3/\text{h}$, alternative flow rate instruments are required.

An extension of the operating range of inlet nozzles below the lower limit by calibration is discussed in Appendix A.1.

Orifice meter

Due to the limitation of the operation range of the inlet nozzles, an orifice meter run conform to DIN EN ISO 5167-2 (2003) has been installed for the measurement of small flow rates. In an orifice meter the pressure difference upstream and downstream in vicinity of a thin orifice plate is measured (Figure 3.4). The flow rate can be determined following equation 3.3:

$$\dot{V} = \frac{C}{\sqrt{1-\beta^4}} e \frac{\pi}{4} d^2 \sqrt{\frac{2\Delta p}{\rho}} \tag{3.3}$$

with the diameter ratio of orifice diameter to pipe diameter $\beta = d/D$ and the flow coefficient C derived from empirical the formula by Reader-Harris/Gallagher (Reader-Harris & Sattary, 1996), which simplifies for the present orifice design to:

$$\begin{aligned} C = 0.5961 + 0.0261\beta^2 - 0.216\beta^8 + 0.000521 \left(\frac{10^6\beta}{Re_D}\right)^{0.7} \\ + (0.0188 + 0.0063A)\,\beta^{3.5} \left(\frac{10^6\beta}{Re_D}\right)^{0.3} \end{aligned} \tag{3.4}$$

In contrast to the inlet nozzles, the orifice meter work based on a fully developed turbulent state of the flow and the corresponding velocity profile. A running length of $\geq 30D$ to the position where the orifice can be installed is required, which demonstrates the major advantage of inlet nozzles due to reduced requirement of space. The turbulent flow state is reliably generated for $Re_D > 5000 - 6000$, corresponding to approximately 5% of the velocity in a comparable inlet nozzle of same pipe diameter. At the same time the differential pressure at the disc is slightly higher than in the equivalent nozzle, which at low flow rates is beneficial with respect to the measurement accuracy of the pressure sensors.

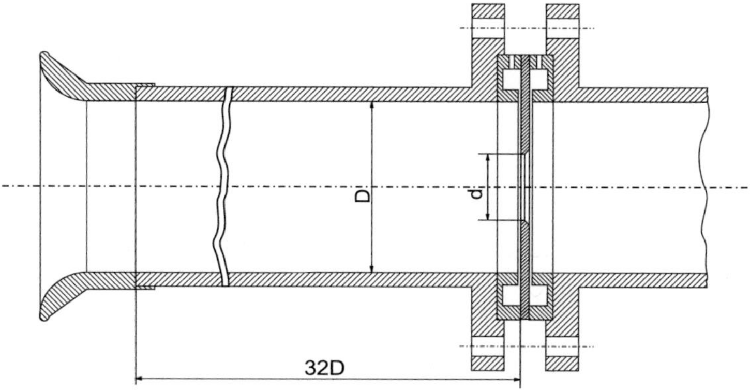

Figure 3.4: Cross sectional view of the orifice meter run with annular chamber type. The connection to the pipe is equipped with centering flanges.

In the present study the orifice meter run consists of an aluminum pipe with a diameter $D = 100.2$mm and two exchangeable disks of $d_B = 60$mm and $d_B = 75$mm. In combination with the available pressure sensors an operation range of $\dot{V} = 40 - 370\text{m}^3/\text{h}$ is covered. The length of the seamless pipe is 4000mm, thereof 3200mm are used for the flow development. To ensure fully developed turbulent flow at smallest flow rates, a zig-zag tripping is installed in direct vicinity to the inlet. The measuring chamber (annular chamber type, Figure 3.4) is custom manufactured (Intra Automation) fitting with an diameter accuracy of tenth of a millimeter to the pipe diameter. The latter is equipped with special centering flanges to precisely align with the chamber. Thus the fabrication tolerances are even significantly better than the specifications according to the norm. An overview of the operation range for both instrument types is given in tables 3.1 and 3.2.

Table 3.1: Operation range of inlet nozzles.

nozzle diameter [mm]	$\dot{V}_{min}$ [m^3/h]	$\dot{V}_{max}$ [m^3/h]
50	190	225
74	225	500
112	470	1150
168	700	2570
262	1620	6250

Table 3.2: Operation range of orifice meters.

orifice plate diameter [mm]	$\dot{V}_{min}$ [m^3/h]	$\dot{V}_{max}$ [m^3/h]
60	38	225
75	68	370

3.1.4 Wind tunnel control

For controlling the flow rate the user can choose between two alternative systems: a LabVIEW-based computer control and a manual analog control. The latter control mainly served for testing during the development of the feed-back control. In both control modes the rotation speed is regulated using the internal control circuit of the frequency converter, which has been custom-programmed for the facility. A personal computer calculates the actual flow quantities: flow rate and velocity in dependence of the connected test section and the type of the flow meter. The computed values are displayed together with the measurement reading of the pressure difference at the flow meter and ambient measurement quantities as well as the computed values for the density ρ and the kinematic viscosity ν.

In the manual mode the rotation speed is adjusted with a potentiometer and the setting is indicated at the display of the frequency converter. Any digital indicator of the computer controlled system can run simultaneously and the full functional range is preserved.

A detailed description of the control software can be found in the thesis of Holderried (2011).

3.2 Test section

The main issue concerning the design of the test section is the choice of its aspect ratio. Various recommendations can be found in literature, as mentioned in Chapter 2.2.2, thus an appropriate compromise was found for $AS = 12$. An aspect ratio of $AS = 20$ would have been desirable, but the enormous difference between width and height, would have required a modified layout of cross section geometry of the settling chamber. This seemed not to be reasonable with respect to the wide range of operation the facility has been projected to run in future projects. The second design objective learned from the literature is the required length to fully develop the flow, especially in the laminar flow case. In this respect the limitation arose from the available space in laboratory: A total length of maximum 4500mm was feasible. A channel height of 25mm has been chosen to be able to realize a total channel length of $150H$.

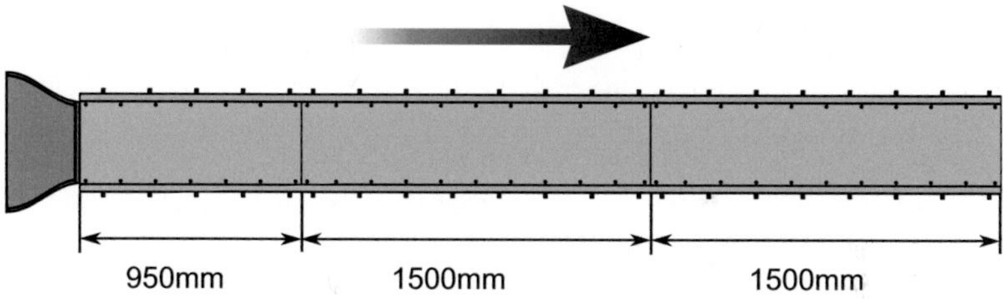

Figure 3.5: Top view of the channel test section with nozzle.

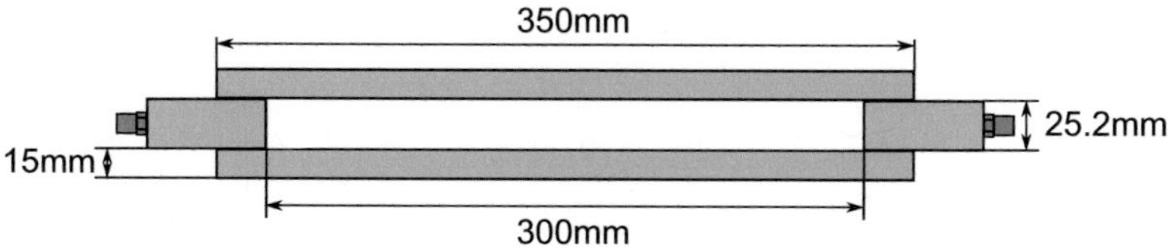

Figure 3.6: Cross sectional view of the channel test section.

Although the channel could have been longer in terms of multiples of the channel height, H was not chosen smaller due to the fabrication tolerances. These considerably affect the measurement accuracy of the flow velocity in the test section for a smaller channel height.

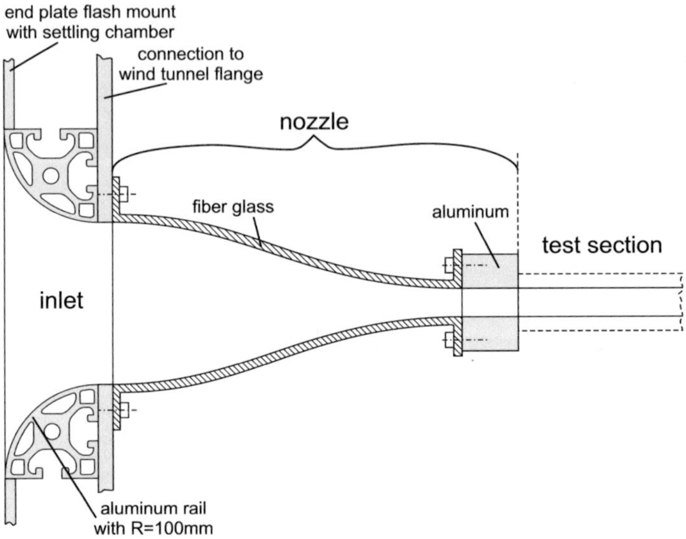

Figure 3.7: Schematic view of the channel inlet and the nozzle.

Care was taken with respect to experimental studies in laminar and transitional flow. Especially the inlet as well as the intersections between channel segments are critical points where improper tolerances can lead to disturbances in the flow. This section gives a detailed review how to realize a continuously smooth surface of highest standards over the complete run length.

Taking into consideration the aforementioned aspects, a construction resulted with two plates for the top and bottom walls as well as two rods for the side walls forming a channel with a width of $W = 300$mm and a height of $H = 25.2$mm corresponding to $AS = 12$ (Fig. 3.6). The channel has a total length of 3950mm. It is connected to the settling chamber by a fiberglass nozzle with a contraction ratio of 6:1. The nozzle has a small contraction ratio of 1.5:1 in spanwise direction in order to be connected to the flat channel at the outlet.

The suction side of the nozzle exhibits an aluminum inlet with quarter cycle shape to ensure a disturbance free entering of the flow into the nozzle (Fig. 3.7). The adjacent nozzle outlet is permanently connected to 40mm long slide of the same geometry and material as the test section. The connection to the test section is therefore quite perfect. The glass fiber part fitted accurately to the air inlet, to the aluminum connector on the test section side with a step of less than 5/100mm. Both intersections have been carefully filled and sanded with an excellent final result. The long-term smoothness of the junction between the the materials aluminum and glass fiber may be be problematic if large temperature differences are encountered. For the present investigation, however, no decrease of quality was been observed.

The test section itself is divided into two parts with respect to the maximum available length of the aluminum for the side walls of 3020mm. The first part has a length of 950mm, the second part of 3000mm. To ensure that all sidewalls are identical in height, all parts were parallely cut from the same plate in direction of grain. The side parts are straight in line without any deflection. To connect the two segments of the test section without any step two guide plates are mounted parallel planar to the side walls as shown in Figure 3.9. This construction turned out to be advantageous: Now, there is a perfect intersection. For changing the cover plates in the long downstream section, the nozzle and the first test section part can remain unchanged. As a result, handling of the test section in case of exchanging the top and bottom plates is relatively easy. To enable replacement, these plates are screwed to the side walls. The cover plates are divided into sections of 950mm, 1500mm and 1500mm as shown in Figure 3.5. To guarantee a smooth surface, the top and bottom walls are made from cast aluminum with a precision milled surface, exhibiting a roughness of $Ra = 0.4\mu$m, while the sidewalls were sanded to equivalent quality (sand grain 1000).

Along the side walls 21 pairs of pressure taps are located which allows a convenient exchange of the top and bottom plates without pressure taps on the test plates. This avoids the difficulty of manufacturing pressure tabs in a structured surface of test plates and guarantees the exact same position and quality of the taps. The first tap is 100mm from the channel entrance and the streamwise distance between the taps is equidistant each 200mm. At the intersections between the cover plates the distance between the neighboring taps is 100mm. This allows to observe eventual misalignment, which would result in a change of the local pressure gradient. It will be shown in Chapter 5 that there is no evidence for a flow disturbance at the intersections. The tabs are positioned at the same streamwise positions on both side walls. Counterpart taps are short cut via a pressure line and yield one averaged reading.

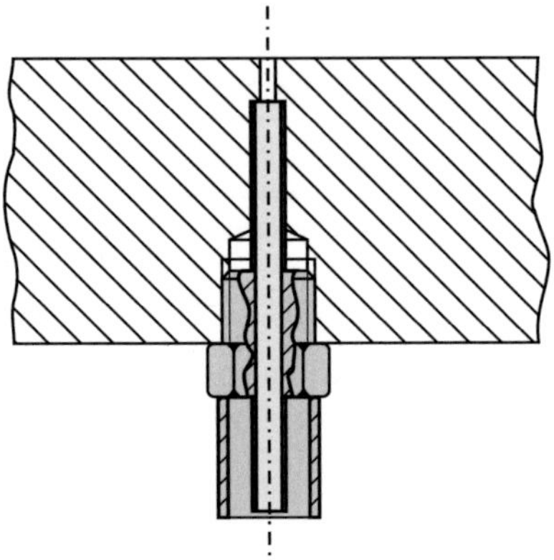

Figure 3.8: Schematic illustration of the construction of the pressure taps.

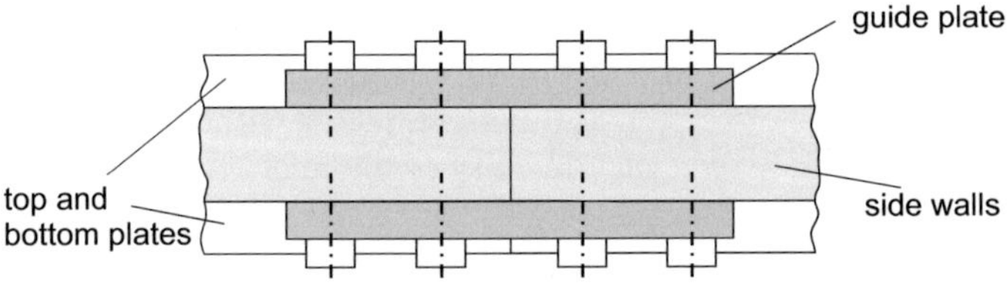

Figure 3.9: Schematic illustration of the connection between segments of the channel section.

The pressure tabs were drilled with a diameter 0.3mm as illustrated in Figure 3.8. Silicon lines of diameter 1.3/0.8mm are connected to the sensors. The connection of the tubes to the pressure tab is implemented with custom-made fittings screwed into the side walls. Part of each fitting is a small 1.5/0.9mm brass tube, transferring the pressure from the tap to the silicon tube. To avoid bending of the protruding brass tubes while handling a protection ring surrounds each of them. In general the construction is in accordance to Nitsche & Brunn (2006).

3.3 Measurement instrumentation

Figure 3.10 illustrates the measurement instrumentation of the system which is split into three main parts:

- Measurement of the ambient parameters
- Measurement of the differential pressure for flow metering
- Measurement of the differential pressure of the pressure taps and of the optionally signals from a hot-wire anemometer.

The recorded ambient parameters contain the temperature T, humidity h and ambient pressure p_{bar}. As mentioned in Section 3.1.3, two differential sensors of type Setra 239D are used to read the differential pressure Δp_f at the flow meter. In the test section a total of 21 pairs of pressure tabs are connected to a mechanical multiplexer (Scanivalve) to transfer the static

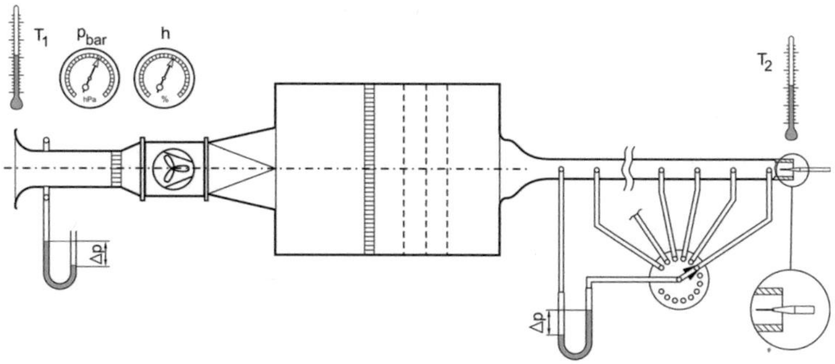

Figure 3.10: Measurement instrumentation.

pressure Δp_l to pressure transducers (MKS Baratron 698A). Optionally a hot-wire can be installed at the tunnel outlet for measuring the turbulence level in the channel.

The interplay of the system is characterized by the collection of measurement data, the communication between components for signal processing and the wind tunnel control as well as the synchronized storage of the data. An overview is presented in Figure 3.11.

The data is acquired by two computers – one for the wind tunnel control and one for the data acquisition at the test section. While the task of the tunnel computer is to deliver the flow rate and the air properties, the test section computer collects the readings of the pressure drop measurement and the hot-wire anemometer. In detail the wind tunnel computer sends the input commands for the rotation speed of the ventilator to the FC. The communication operates via AD/DA-converter (National Instruments (NI) USB-6211) to activate the internal controller of the FC. The same device is utilized to digitize the readings of three PT100 resistance thermometers reporting the temperature at the inlet of the orifice meter, at the inlet of the nozzle and at the outlet of the tunnel. Additionally the differential pressure Δp_f from the Setra Sensors of the flow meter is recorded. Two humidity sensors (Sensirion SHT2) are directly connected to the PC via USB and the ambient pressure p_{bar} sources from manual input (source: meteorological station of Hessisches Landesamt für Umwelt und Geologie (HLUG)). From this data collection the air quantities ρ and η are computed, which are needed for the accurate calculation of the flow quantities Re_b, $\dot{V}$ and U_b. Via network communication these quantities are send to the measurement computer.

The measurement computer simultaneously acquires data, receives a continuous data flow from the wind tunnel control and synchronously saves these data on the hard disk. This computer is located at the test section and controls the Scanivalve multiplexer via an AD/DA-converter (NI-6009) to switch between the positions of the pressure taps connected of the pressure transducers. In the opposite direction position report of the Scanivalve is received. Two parallely driven Scanivalve multiplexers are used, each connected to a highly accurate capacitive pressure sensors (MKS Baratron 698A), to obtain simultaneous pressure measurements at two streamwise tap positions. The pressure readings are directly digitized by the Baratron systems and sent to the computer using a serial port (RS232). If optionally a hot-wire is installed at the tunnel outlet, the computer directly communicates with the CTA-device (Dantec Streamline). The analog output of the CTA bridge is digitized with a third AD-converter (NI-X-6361) enabling high sampling rate.

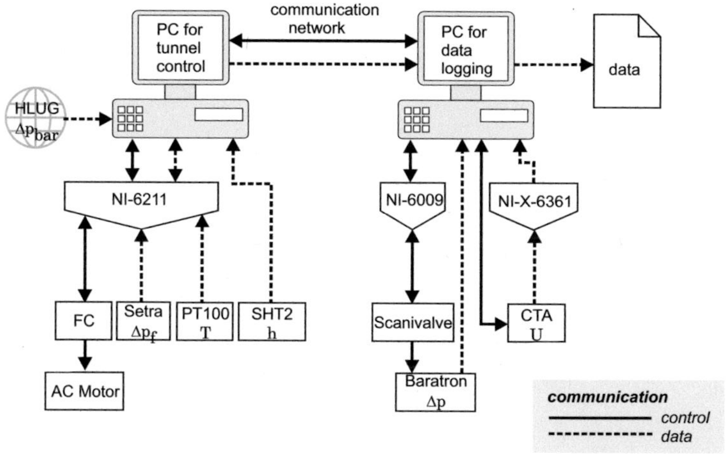

Figure 3.11: Organigram including signal processing, control and communication of the system.

4 Measurement uncertainty

The main issue of the uncertainty estimation is to state the measurement accuracy which can theoretically be achieved with the present facility. In the following Chapters we will recheck whether measurement data confirms the estimated accuracy.

Before the calculated uncertainty is presented, the procedure is explained to clearly define which conclusions can be drawn from the specified errors. This is elementary with regard to the comparability to literature data, where the uncertainty estimation may be conducted under different assumptions and methods. Thus a direct comparability of the magnitude of errors unfortunately cannot be ensured. However, a comprehensive error analysis as presented here can hardly be found in similar investigations.

The definitions orientate at the books of Tropea *et al.* (2003) and Coleman & Steele (2009).

4.1 Modus operandi

The uncertainty calculation identifies and analyzes the sources of errors and subsequently combines the individual estimates to an overall result. The analysis considers the quantities $\dot{V}$, Re_b, c_f and $\tau_w/\tau_{w,0}$, which were deduced as a fundamental base for the comparison of the skin-friction for different surface conditions in the theoretical considerations (Chapter 2). First static errors are considered theoretically, i.e. errors which are independent of time. Hence the calculated deflections are always referred to the "true value", which is an ideal quantity. For the individual errors of the measurement devices we assume that the deviation between the true value and the right value is negligible and the two quantities can be equated. The right value represents the reference value for the calibration certificates of the devices specified by the producer. Time dependent or dynamic errors $\delta = f(t)$ are determined from measurement data and added to the result.

When error limits are plotted in the measurement data, the uncertainty is normally given in terms of absolute values. Strictly regarded, the error intervals would need to be sketched into diagrams referring to the true value. Since the true or right value is generally not available for the determined quantities the reference point for the error intervals is unidentified, too. Hence according to the current method, specifications of errors refer to measurement value. Alternatively error ranges are sketched around a fit-line through scattered data. The latter processing strategy avoids to map error bars for any single data point and is owed to the clearness of the plotted data.

As the uncertainty analysis will show, in the present case most of the errors are systematic and not random. An examination based on the Gaussian error propagation law is not appropriate. In consequence the uncertainty estimation is conducted using linear error propagation (Eq. (4.1)). The calculation delivers the maximum error of a single data point for the case of addition of all errors δx_n with maximum magnitude and same sign. The resulting error limits define the potential maximum measurement error δy_{max}. This does not necessarily mean that an error of such amount ever occurs in the measurement. Additionally for the flow rate a 95%-confidence interval according to DIN EN ISO 5167-1 (2003) is specified.

$$\delta y_{max} = \sum_{i=1}^{n} \left| \frac{\partial y}{\partial x_n} \right| \cdot \delta x_n \tag{4.1}$$

Besides the maximum uncertainty from the linear error propagation, the uncertainty of the reproduction is calculated according to the same approach. The uncertainty is separated into a systematic (bias) and a random (statistical) fraction. Systematic errors, which influence each measurement with same magnitude and sign, do not affect the reproducibility of the measurement (Fig. 4.1). In consequence the accuracy is consulted for the comparison with literature data, while the precision is fundamental for the calculation of the uncertainty of the skin-friction changes $\delta\tau_w/\tau_{w,0}$.

The resulting uncertainty limits depict the range of scatter, in which all measurement data need to reside, provided that the uncertainty estimation is neither based on inapplicable assumptions nor potential error sources are excluded.

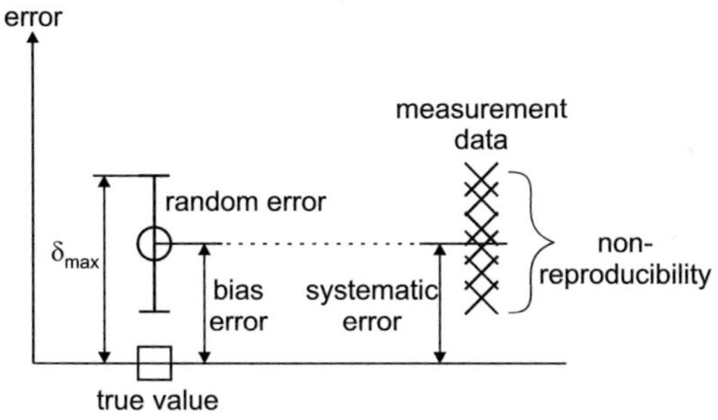

Figure 4.1: Definitions for different kinds of errors.

In the present study measurement uncertainties are resulting from three sources:

- the flow rate measurement, with the acquisition of ambient quantities as its elemental component,
- the accuracy of the determination of the pressure difference along the test section,
- and the steadiness of the flow during the measurement.

The discussion will follow these main aspects. Successively the individual error sources are examined to demonstrate their influence on the characteristic quantities $\dot{V}$, Re_b, c_f and $\tau_w/\tau_{w,0}$.

One difficulty concerning the uncertainty estimation was the inconsistent specification of the measurement uncertainty by the producers of the different sensors and devices. Examplarily Sensirion specified a maximum error, while Setra gave maximum errors for the single sources, but specified a combined accuracy for the following Gaussian error propagation. While all devices which are according to normed standards, like temperature probes, are specified with a 2σ-confidence interval. In the following the approach is made to calculate the maximum uncertainty for all devices to achieve a consistent uncertainty propagation. As the true value is, the maximum error, of course, represents an ideal quantity and its determination is only an approximation. The result of the present uncertainty estimation corresponds approximately to a 3σ-confidence interval, which describes 99.7% confidence.

4.2 Flow rate measurement

Uncertainties of the flow rate measurement refer to four sources:

- acquisition of ambient parameters,
- accuracy of pressure sensors,
- uncertainty of the calibration parameters,
- and fabrication tolerances.

These issues are successively discussed in the following paragraphs and subsequently merged to a total uncertainty of the flow rate.

Since the facility has been put into operation, the measurement instrumentation has been constantly revised and upgraded. During the project a continuous improvement of the measurement accuracy has been achieved from improved accuracy of determination of the ambient data. After the first measurement campaign (micro grooves) the air humidity was taken into account. Additionally the temperature measurement has been substantially improved. Hence, the measurement accuracy of ambient parameters differs between the test campaigns. While initially the uncertainty is discussed for the lastest step of development, the evolution of the facility is taken into account for individual calculation of measurement uncertainty of a measurement campaign and pointed out for the corresponding investigation.

4.2.1 Ambient quantities

The determination of the ambient conditions of air density ρ and dynamic viscosity η is fundamental to accurately calculate the flow rate $\dot{V}$ as well as the corresponding flow velocity U_b and the Re_b in the test section. As introduced in Chapter 2.3 the air density is calculated with the ideal gas law depending on the ambient pressure and temperature. In addition, the air humidity is taken into account by employing the empiric Magnus formula (Eq. (2.52)). The viscosity determination is based on Sutherland's formula (Eq. (2.54)).

For the calculation it is assumed that these empirical formulas are exact. In the narrow range of temperature between $14 - 25°C$ in the laboratory combined with the small deviation from the reference value $T_n = 15°C$ this assumption marks a marginal factor.

The uncertainty focuses primarily on the state of the art of the facility including the detection of air humidity. The additional errors induced by neglecting the air humidity and less accurate temperature detection at the beginning of the study are given afterwards.

The ambient temperature is determined by three PT100 resistance (accuracy class 1/3B) thermometers at different positions in the laboratory. The sensors are located at the inlet of orifice meter, at the suction side of the inlet nozzles and at the outlet of the channel. The measurement error is basically systematic due to a zero deviation in the considered temperature range (Eq. (4.2)). Through simultaneous calibration of all three sensors in an ice water mixture the absolute error has been minimized to $< 0.15°C$. The uncertainty calculations are based on a value of accuracy of the temperature detection of $\delta T = \pm 0.2°C$ including some assumptions introduced in Section 4.2.6.

$$\delta T_{(1/3DINB)} = \pm(0.10 + 0.00167T) \tag{4.2}$$

The humidity is recorded at two positions in the laboratory with two Sensirion 2SHT-sensors having an accuracy of $\pm 3\%$.

The barometric pressure is obtained from the meteorological station of Hessisches Landesamt für Umwelt und Geologie (HLUG) in the Rudolf-Müller-Anlage originally converted for sea level (p_{asl}). Hence data needed to be recalculated for an absolute altitude of $z_m = 170\text{m}$ for the laboratory, which has been done based on the internal procedure of the station:

$$p_{asl} = \frac{z_m \, 1.2\text{hPa}}{10\text{m}} + p_{z_m} \tag{4.3}$$

In fact this formula is a rough estimate to convert to the pressure at sea level since it neglects the influence of the temperature on the air density. However it does not affect the the uncertainty of the present investigation. Since the formula is utilized to recalculate the original data of the measuring station, the received value for the station's absolute altitude will be very accurate. Concerning the level of the laboratory being only 14m above the meteorologic measurement station, the arising error is in a range of few Pascal and thus negligible. An accuracy of ±1hPa can be estimated. Table 4.1 gives an overview of the uncertainty for determination of ambient parameters.

Table 4.1: Uncertainty ambient data

δT[°C]	δp_{bar} [hPa]	δh [%]	$\delta \rho$ [%]	$\delta \eta$ [%]
0.2	1	3	0.19	0.05

Considering the uncertainty of the determination of the ambient data at the initial state of the facility, the lower accuracy of the temperature detection and neglecting of the humidity degrades the accuracy. The temperature has been read with a PT-resistor from Dantec Streamline with a resolution of 0.1°C. The accuracy of this device is not specified by Dantec, therefore its zero offset is estimated by a comparison against the calibrated temperature probes of the actual set-up to be $\delta T_{zero} = -0.5$°C. Regarding the humidity data from the later measurement campaigns a range of $h = 25 - 75\%$ can be deduced. If we consider the humidity is varying in that range, an error of $0.1 - 0.5\%$ arises. Contrary to the temperature, where a systematic deviation is found, the neglect of the humidity leads to a scatter of ±0.2 around a mean error of 0.3%.

Table 4.2: Uncertainty ambient data for initial set-up

δT[°C]	δp_{bar} [hPa]	h [%]	$\delta \rho$ [%]
0.4	1	25	0.3
0.4	1	75	0.7

4.2.2 Uncertainty of the differential pressure measurement for flow rate determination

The differential pressure at the nozzles or the orifice meter is measured by two capacitive sensors of type Setra 239D. If the range of the first sensor of 125Pa is exceeded, the other Setra with a range of 625Pa is read. Both sensors feature an accuracy of 0.07%FS (of full scale) (detailed specification c.f. table 4.3). The error specification is the combined error based on a Gaussian error propagation of the maximum errors originating from different sources. This uncertainty specification corresponds to a certain confidence interval and therefore is not consistent with the present approach to estimate the maximum uncertainty. Figure 4.2 visualizes how the approximation for the maximum errors of the sensors were calculated. In addition to the specified measurement accuracy a hysteresis error as well as the deviation resulting from the zero point deviation was added. The zero point deviation of $< \pm 0.1$Pa and ± 0.5Pa, corresponding to $< 0.125\%$ FS, was obtained in the measurement campaigns for both sensors. Last but not least it needs to be considered that the sensor's accuracy may degrade with time. According to the specifications of the manufacturer, for the period of the investigation the uncertainty calculation includes a drift of the span of 0.2 and 1Pa corresponding to three times the specified accuracy.

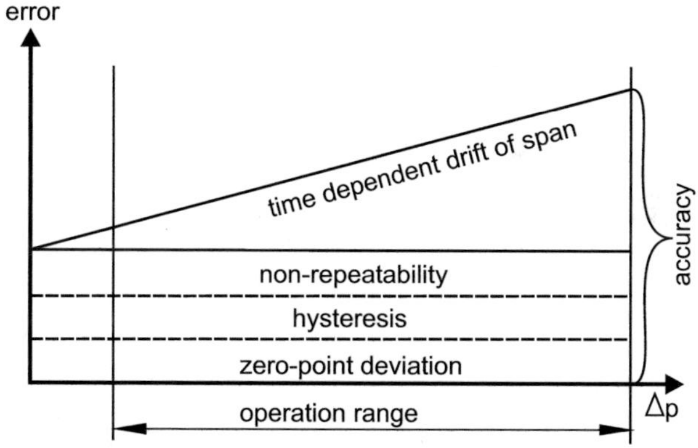

Figure 4.2: Combined uncertainty of the pressure transducers Setra 239D.

Before the first measurement campaign the differential pressure sensors for the flow rate determination and the pressure sensors for the detection of the pressure loss along the test section, which exhibit an even higher accuracy (compare Tab. 4.5), were balanced to each other. The comparison has been rechecked twice a year. A relative deviation between the sensors could not be discovered ($< 0.1\%$). The minimal deviations between the sensors remained constant over the time of the study. Since it has not

Table 4.3: Product specifications Setra 239D

Setra model 239D	
accuracy RSS at constant temp.	±0.07% FS
non-linearity BFSL	±0.10% FS
hysteresis	0.10% FS
non-repeatability	0.02% FS
compensated range °C	−1 to +65

been possible to review the absolute errors by calibration, regardless an uncertainty due to time dependent drift of the span was added.

4.2.3 Fabrication tolerances

Fabrication tolerances play a role for nozzles and orifices, but different parameters of the geometry are critical. In case of the nozzles the main parameters are exact manufacturing of the inner diameter, the curvature at the inlet and an accurate drilling of the pressure tab. Regarding the orifice meter, the disk itself can be purchased in high grade finish. In this case the pipe is the inaccurate element, where the centering of the flanges of the intersections is important. In the present study the challenging aspect were treated with special care, which is described in detail in Chapter 3.1.3. Fabrication tolerances are summarized in table 4.4.

At least the results for the 50mm and 74mm inlet nozzles were deviating from expectations. The nozzles run with perfect reproducibility, but show a systematic deviation of the flow rate. This became obvious in the reference measurement for the channel in comparison to literature data. Therefore the nozzles were calibrated with reference to the orifice meters (described in Appendix A.1) and thus the accuracy is the same for both types of instruments. It is assumed that the deviation of the volume flux measured by the nozzles results from an inaccurate fabrication of the inlet radius of the instruments. This is inconvenient regarding that the nozzles should nominally provide higher absolute accuracy than the orifice meters (see Sec. 4.3).

Table 4.4: Manufacturing tolerances for the orifice meter run with the opening of orifice disk d and the pipe diameter D

d [mm]	δd [mm]	D [mm]	δD [mm]	β	$\delta\beta$
60	0.005	100.2	0.05	0.599	0.0003

4.2.4 Uncertainty of the calibration parameters

Uncertainties of the DIN-specified flow coefficients mainly result from the fact that the flow rate instruments are not calibrated in the particular installation situation. The norm specifies frame conditions for fabrication tolerances as well as conditions for the installation. Especially in the case of the orifice meter run the uncertainty of the given calibration parameters must include quite large tolerated deviations. In consequence the error for a certain flow instrument is primarily systematic, which implies that the reproducibility of the flow instruments is much better than the absolute accuracy.

Uncertainty of the calibration parameters includes the uncertainty of the flow coefficient α (nozzle) or the orifice coefficient C (orifice) (Eq.(3.2) and Eq. (3.3) in Chap. 3). Both have a major impact on the absolute measurement accuracy. The calculation is based on empirical formula that can again be found in the engineer standards (DIN24167-2 (nozzle) and DIN EN ISO 5167-1 (2003) (orifice)). In the present study the 50mm nozzle has been calibrated to work at smaller flow rates. This tedious work and its unforeseen difficulties are described in Appendix A.1.

4.2.5 Combined uncertainty of flow rate measurement

The individual error sources are now added to a combined uncertainty of the flow rate measurement. The combined uncertainty will be presented as the maximum error for the standard conditions defined in Table 2.1. In this way the estimate for the flow rate is equal to the uncertainty for the mass flow $\dot{m}$, since the weighting of the density, which enters as square root in denominator for $\dot{V}$ and in nominator for $\dot{m}$, is approximately the same for the small uncertainty of the density calculation.

Figures 4.3 and 4.4 show the uncertainty of the flow rate for the nozzles and for the orifice meters for different D or d, respectively. For both types of flow meters systematic errors are dominating, which mainly arise from the uncertainty of the flow coefficients and are larger for the orifice meter. The non-reproducibility ϵ is considerably better and almost identical for both types of flow instruments. For the uncertainty as well as the non-reproducibility of the flow rate a peak originating from the switching between the Setra sensors with different measurement ranges is clearly visible. One can deduce that the accuracy of the pressure transducers is of main impact for the reproducibility.

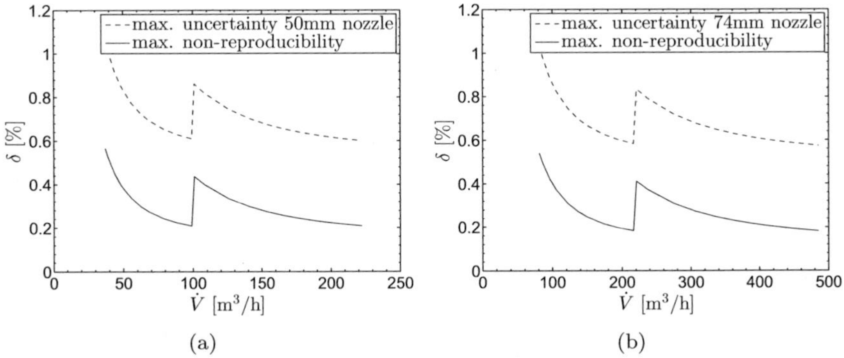

Figure 4.3: Max. error of flow rate for nozzles with $D = 50\text{mm}$ (a) and $D = 74\text{mm}$ (b).

As mentioned before, the fabricated nozzles did not comply with the intended geometry and were calibrated in reference to the orifice meters. Thus, the following considerations focus on the orifice meters.

Alternatively, a calculation of the uncertainty for $\dot{m}$ can be derived from the norm DIN EN ISO 5167-1 (2003) for a 2σ confidence interval.

$$\frac{\delta \dot{m}}{\dot{m}} = \sqrt{\left(\frac{\delta C}{C}\right)^2 + \left(\frac{\delta e}{e}\right)^2 + \left(\frac{2\beta^4}{1-\beta^4}\right)^2 \left(\frac{\delta D}{D}\right)^2 + \left(\frac{2}{1-\beta^4}\right)^2 \left(\frac{\delta d}{d}\right)^2 + \frac{1}{4}\left(\frac{\delta \Delta p}{\Delta p}\right)^2 + \frac{1}{4}\left(\frac{\delta \rho}{\rho}\right)^2}. \tag{4.4}$$

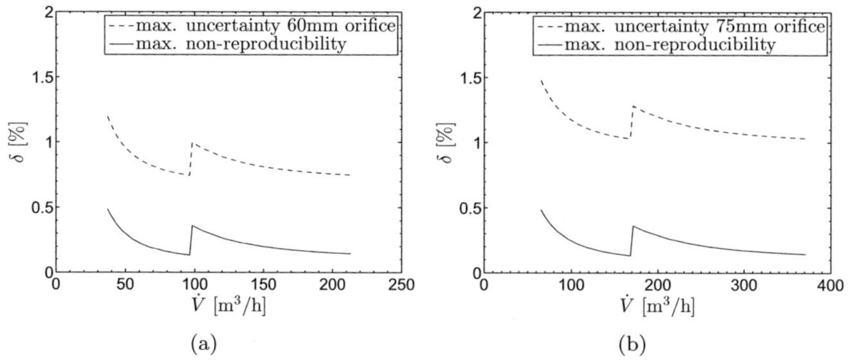

Figure 4.4: Max. error of flow rate for orifice disks with $D = 60$mm (a) and $D = 75$mm (b).

This procedure is neither according to a Gaussian nor to a linear determination of error propagation, since the calculation following DIN EN ISO 5167-1 (2003) equally handles statistical and systematic errors by accumulating the modulus using a weighting of the single error contributions. In particular, the weighting of the uncertainty arising from differential pressure measurement is not plausible to the author. It leads to a very small influence of the pressure measurement on the combined uncertainty as exemplarily shown in Figure 4.5. Thus, instead of using the uncertainty consideration according to DIN, linear error propagation is used for the estimation of uncertainties. Nonetheless, in the present case the procedure according to DIN delivers approximately the same result as the linear error propagation.

4.2.6 Uncertainty of Reynolds number in the test section

If we consider the uncertainty of the Reynolds number in the test section based on equation (4.5) and insert the flow rate, four potential error sources affecting Re_b are identified:

- viscosity,
- tolerances of the test section geometry,

- temperature change between the flow meters at the inlet of the wind tunnel and the test section,
- differences between static pressure in the test section and the ambient pressure.

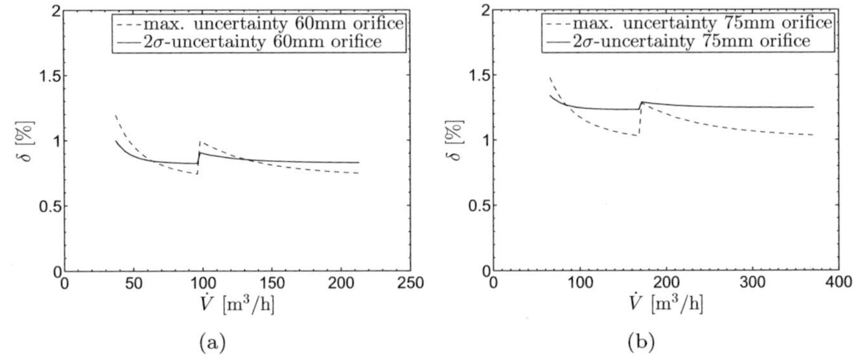

Figure 4.5: Uncertainty estimation of flow rate according to DIN in comparison to linear error propagation for orifice disks with $D = 60\text{mm}$ (a) and $D = 75\text{mm}$ (b).

The influence of the uncertainty of the viscosity on the flow rate is marginal, since only the flow coefficient C, which is nearly constant for the range of Re_D, is weakly depending on ν. In contrast Re_b is proportional to $1/\nu$. From the additional error sources $\delta\nu$ is the only one with statistical character.

$$Re_b = \frac{U_b\, H}{\nu} = e\, C \frac{A_D}{A_C} \frac{\sqrt{\frac{2\Delta p}{\rho}}\; H\, \rho}{\eta} = e\, C \frac{A_D}{W H} \frac{\sqrt{2\Delta p}\, H \sqrt{\rho}}{\eta} \tag{4.5}$$

Fabrication tolerances are represented by the ratio between the cross sectional area of the nozzle and the test section A_D/A_C. The additional error, compared to $\dot{V}$, is very small, since the relative tolerance of the channel width $\delta W/W$ is small and the channel height H cancels out (see (4.5)).

Temperature changes between inlet and outlet are originating from heat up of air passing through the facility: heat emission from the motor and dissipative losses (skin-friction, Borda-Carnot losses). All dissipative losses lead to a temperature increase of the fluid with the exception of the Borda-Carnot loss at the outlet. The arising pressure drop losses need to be compensated by the ventilator, which is producing additional heat emission in dependence of the power consumption and the efficiency. A theoretical calculation including the previously mentioned heat sources, indicated very small influence for small Re_b. The heat-up of the fluid has been computed to estimate the difference between inlet and outlet temperature ΔT shown in Figure 4.6. To confirm the theoretical consideration, measurement data from the temperature difference between inlet and outlet for stationary operating mode are added to the plot. Measurement and computation for the 74mm inlet nozzle are in very good agreement. If we use the mean value of inlet and outlet temperature for the calculation of $\dot{V}$ as well as of Re_b, the resulting error in the temperature is smaller than the error δT we estimated for low Re_b. For $Re_b > 18000$, we have a low impact and need to determine $\dot{V}$ and Re_b based on different ambient temperatures. In the present case the temperature increase within the test section is low compared to the other heat sources of the system ($< 15\%$). Thus the temperature in the test section can be regarded constant. The validity of this assumption is also limited to the moderate Re_b of the present study.

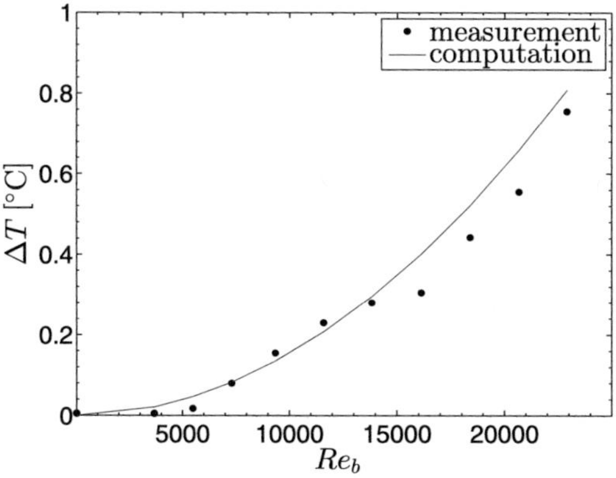

Figure 4.6: Difference between inlet and outlet temperature (calculation of Re_b based on mean temperature).

Regarding the static pressure at the flow rate instruments, the marginal pressure loss to the location of the pressure tab is automatically included in the DIN-calibration for the nozzles. For the orifice meters the static pressure in the pipe in the proximity of the orifice disk is almost equal to the barometric pressure due to the comparably low velocities in the pipe (restricted by the range of the Setra sensors $\Delta p_{max} = 625$Pa).

The static pressure at the channel inlet is the overpressure of the settling chamber, which is driving the flow. The pressure decreases due to energy dissipation along the test section and is equal to the barometric $p = p_{bar}$ at the outlet of the test section. Figure 4.7 shows the pressure difference from channel inlet to the outlet based on the assumption of the validity Dean's law for the whole channel. If we consider the overpressure for the complete test section, this effect is definitely negligible for $Re_b < 23000$. In this operational range the changes of temperature and static pressure are negligible small in the channel, which is essential for the validity of the criterion of constant density. If the channel section works at $Re_b > 30000$ these effects will become relevant. Note that in this case Re_b would slightly change along the channel.

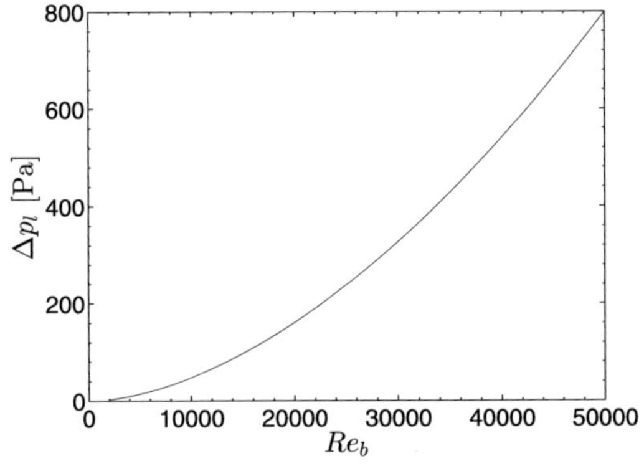

Figure 4.7: Estimation of the pressure loss in the test section according to Dean's law.

As a result of the preceding analysis, the error graph for Re_b in Figure 4.8 is very similar to the one for $\dot{V}$, since the viscosity and the fabrication tolerances bring very small additional errors.

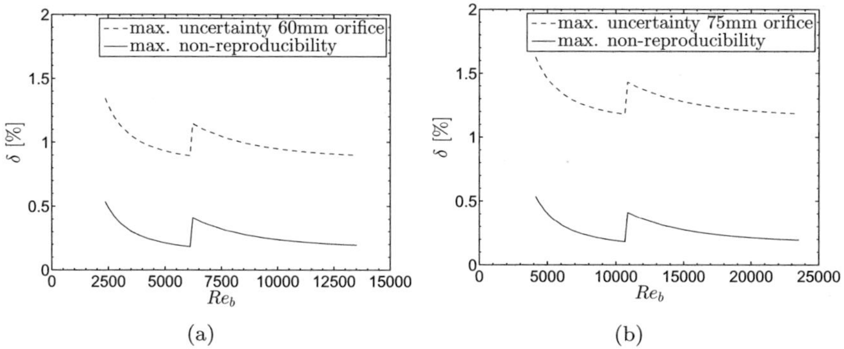

Figure 4.8: Max. error of Re_b for orifice disks with $D = 60\text{mm}$ (a) and $D = 75\text{mm}$ (b).

4.3 Determination of pressure drop along the test section

For the measurement of the pressure evolution along the test section, two error sources need to be considered: on the one hand a possible deviation originating from the pressure taps and on the other hand the accuracy of the differential pressure sensors.

While common pressure sensors exhibit an accuracy of $\pm 0.1 - 0.5$ % FS, the two Baratron 696A sensors offer $\pm 0.12\%$ R (of reading) in three measurement ranges. Even smallest values can be detected with the highest accuracy. In order to obtain the most accurate results possible, the measurement range of the sensor needs to be adjusted to the expected pressure losses. Baratron sensors are temperature compensated and after a certain heat-up time of a few days show minimal dependency of the zero point on the ambient temperature. The drift of the zero point is compensated manually with an accuracy of ± 0.01Pa for all ranges. It has been done so and minuted for every measurement day.

Disadvantages of the Baratron sensors should not be omitted here. The low temporal resolution of 10Hz limits the application to more or less stationary flows. Concerning the present work this is not a constraint, since dynamic processes are not examined using pressure measurement technique. Detailed specifications of the Baratron sensors are found in Table 4.5.

Table 4.5: Specifivations of differential pressure sensors MKS Baratron 698A

Specifications of type	698A
accuracy	0.12% of reading ± zero/Span coeff.
full scale pressure ranges (Torr)	$1, 10$
resolution (FS)	1×10^{-6}
sensor temperature	regulated at 45°C
temperature effects on span	$<$ 0.002% R / °C (20ppm)
temperature effects on zero	$<$ 15 PPM, FS /°C
type of measurement	differential
useable resolution	5 decades

The producer MKS calibrates the devices to less than half the error of a 3σ-confidence interval, which is proven by a calibration certificate. This specification is assumed to be in accordance to the present uncertainty calculation.

The pressure taps were analyzed following Tropea *et al.* (2007) with respect to potential errors originating from:

- the tap diameter,
- the relative tapping depth,
- the condition of the tapping orifice.

The tap diameter is of major importance, since all other potential errors are highly dependent on this parameter. Therefore the tap diameter was chosen small with only 0.3mm and taps were manufactured in high quality. If the tap diameter is normalized by the viscous length scale δ_ν, we obtain $d^+_{tap} < 15$ for the riblet study. For all other studies it is considerably smaller ($d^+_{tap} < 10$). Due to the small diameter, the errors are negligible ($\delta \ll 0.1\%$) with respect to all mentioned aspects.

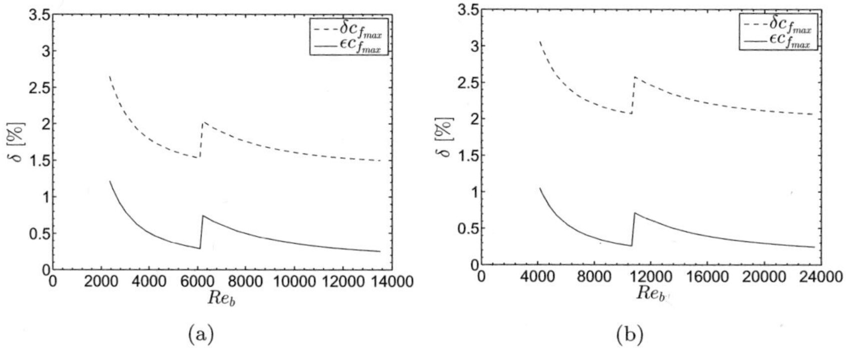

Figure 4.9: Max. error and non-reproducibility of c_f for orifice disks with $D = 60\text{mm}$ (a) and $D = 75\text{mm}$ (b).

4.3.1 Uncertainty of friction coefficient

The only further source of error affecting the friction coefficient is the pressure drop along the test section as can be obtained from:

$$
\begin{aligned}
c_f = \frac{\Delta p_l H}{\rho_c\, U_b^2 L} &= \frac{\Delta p_l H}{\rho_c (e\, C \frac{A_D}{A_C} \sqrt{\frac{2\Delta p_f}{\rho_f}})^2 L} \\
\text{for } \rho_c = \rho_f = \rho: \; c_f &= \frac{\Delta p_l H}{(e\, C \frac{A_D}{A_C})^2 2\Delta p_f L}.
\end{aligned}
\tag{4.6}
$$

As shown in the previous section the uncertainty of the pressure drop is small compared to the uncertainty of the flow rate. Thus the main impact results from the quadratic relation on U_b. The influence of fabrication tolerances grows, since now the channel height H does not vanish. For example a deviation of 0.25mm would lead to an error of 1% in U_b. However, the fabrication tolerances are fare less for the present set-up (cf. Chapter 3.2) If differences in the density between the location of flow rate meter ρ_f and the channel section ρ_c - basically induced by temperature increase and large overpressure in the test section at high Re_b - are neglected, a systematic linear error dependency is obtained. Note that as long as the assumption of a nearly constant density through the facility holds, c_f does not depend on ρ (see Eq. (4.6).

Assuming a quadratic relation between the flow velocity U_b and the pressure loss in the channel, a variation of the flow rate has twice the impact on the pressure loss. If this is transferred to a plot of c_f versus Re_b, the error of the flow rate determination on the abscissa affects the values on the ordinate of the c_f-value by factor two.

In Figure 4.9 the bias and random errors are distinguished. It is clearly visible that a large portion relates to systematic errors. Thus reproducibility of c_f is fairly good. Figure 4.10 splits the total uncertainty of the friction coefficient into the individual error sources: the pressure sensors for Δp_f and Δp_l, C, β and others. The major part of the uncertainty results from the flow coefficient C, which represents a pure systematic uncertainty. The larger absolute uncertainty of the 75mm disk is a result of δC increasing with β. Obviously a calibration of the flow meter could reduce the absolute uncertainty by one half.

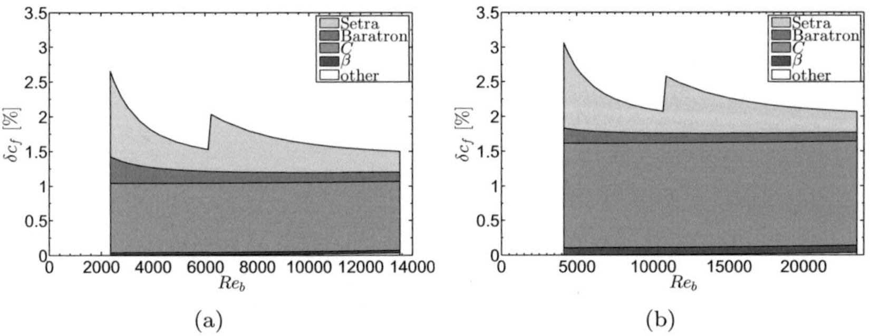

Figure 4.10: Max. error of c_f split into error sources for orifice disks with $D = 60\text{mm}$ (a) and $D = 75\text{mm}$ (b).

In literature the uncertainty of the instrument for skin-friction detection is commonly given, while the effect of flow rate uncertainty is not taken into account, although its impact on the data is generally larger than the error of the pressure measurement.

4.3.2 Determination of skin friction changes in dependence of the measurement strategy

The experimental set-up has to provide an accurate information about the difference in skin-friction between a smooth reference and test case. For that purpose the absolute accuracy is not the limiting condition but the reproducibility, since systematic errors in the measurement do not alter the ratio $\tau_w/\tau_{w,0} = c_f/c_{f,0}$ (Fig. 4.11 (1) and (2)). Thus the non-reproducibility of the pressure sensors for flow rate as well as for detection of pressure drop along the channel and the sensors for the ambient parameters are the crucial issues. The largest identified source of uncertainty was the one related to the flow coefficient C, which does not affect the result. Nevertheless, the flow rate contributes the largest error to the measurement as illustrated by the horizontal error bar in Figure 4.11. If in the worst case the computation of $\tau_w/\tau_{w,0}$ is based on two data points, one for test surface and one for the reference, the non-reproducibilities add and result in quite large uncertainty. The confidence can be improved, if a test series is repeated several times and then averaged, as done in the present study (Fig. 4.11 (3)). Then reproduction errors are statistically distributed and a Gaussian distribution can be assumed. The error propagation law according to Gauß can be applied. The average of the data and a confidence interval σ is computed (4). If we employ a confidence interval of 1σ, the uncertainty for a comparison of the test and reference measurement can be determined. Expecting identical standard deviation for both cases, the uncertainty will propagate as:

$$\delta_{1\sigma} = \sqrt{\sigma_0^2 + \sigma_{test}^2} = \sqrt{2}\sigma. \tag{4.7}$$

Two factors can improve the accuracy of the measurement: The first is the time dependence of the reproduction error. Part of the previously discussed errors originates from deviations of the sensors with time, which can also be random. For example the zero point deviation of the Setra sensors did not show a specific preference in time, but varied in both directions. In a measurement campaign which took weeks or months, the deviation was observed to be statistically distributed, but was approximately constant within one day. If a test series can be conducted within one day the data scatter is be minimized and the accuracy of $\tau_w/\tau_{w,0}$ is increased.

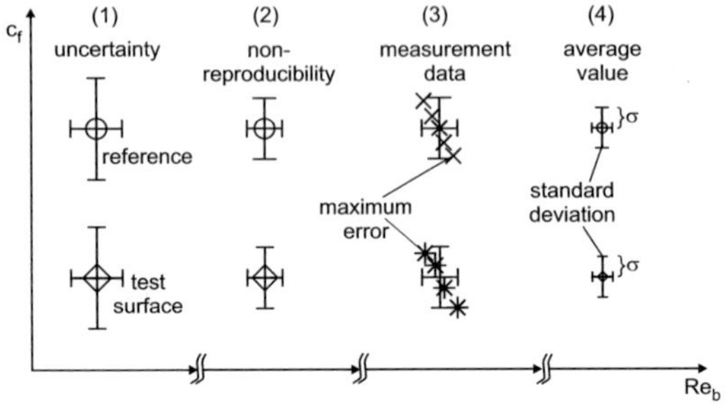

Figure 4.11: Influence of the uncertainty on the significance of the data.

The second and essential factor is the strategy of testing. Figure 4.12 schematically visualized the maximum error occurring from the the strategy on base of a single measurement. The major impact on non-reproducibility ϵ of c_f originates from the flow rate measurement, which is symbolized by the horizontal error bar, since ϵc_f is quadratically proportional to U_b.

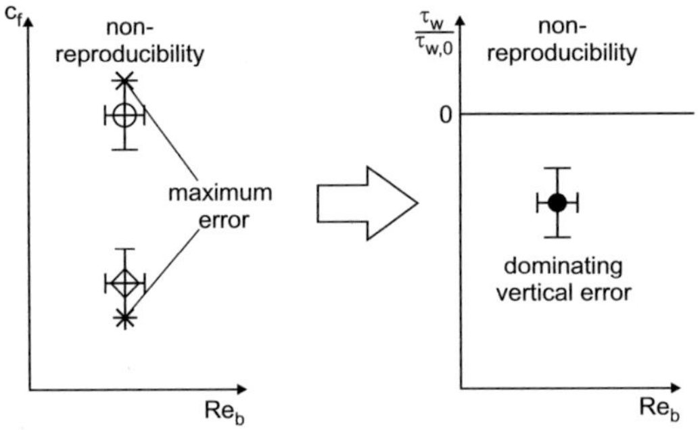

Figure 4.12: Strategy 1: comparison of reference and test case from separate measurements.

As Figure 4.13 illustrates, the situation changes if the measurement of test and reference conditions can be conducted simultaneously. For example in a long channel there can be two segments, one with a smooth reference surface and one with a modified surface condition. In this case the flow rate must definitely be same for both surface conditions. All errors arising from the flow rate posses the same magnitude and sign. The consequence is a low influence of the flow rate uncertainty to the amount of $\tau_w/\tau_{w,0}$. Nonetheless scatter will be visible in the data, when repeating the measurement. In fact $\tau_w/\tau_{w,0}$ is quite exactly known, but the horizontal position of a data point is still affected by the non-reproducibility of the flow rate.

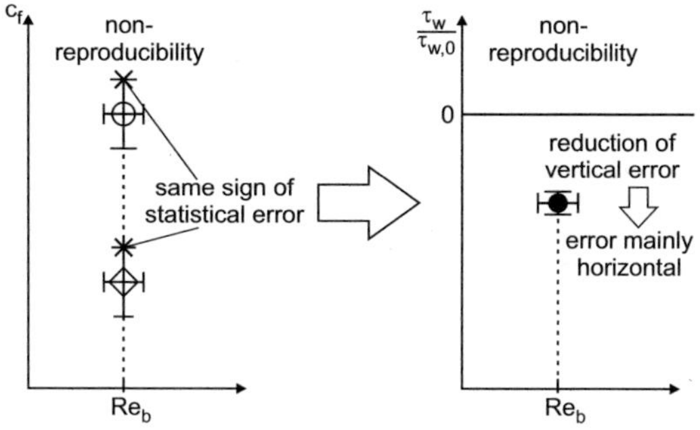

Figure 4.13: Strategy 2: comparison of reference and test case within the same test run.

During the measurement campaigns different strategies will be utilized to gain the most accurate result in the particular constellation. In general it is beneficial to enable a direct comparison to the reference. In most cases measurement of reference and test case are conducted at different times. The time inbetween leads to larger scatter in the data, since influence of deviations in the sensors is present. The present uncertainty estimation includes these aspect for $\dot{V}$, Re_b and c_f. Although the absolute accuracy is stated to be the same for all studies, the non-reproducibility will differ significantly depending on the measurement strategy, because the measurement strategy essentially influences the magnitude of the statistical errors.

Changes in the ambient temperature, pressure or humidity are small considering a single measurement or even one day of tests. Temperature changes were smaller than 0.3°C per hour, maximum variation in the barometric pressure has been $\Delta p_{max} < 1$ hPa/h. If a test run did not satisfy these criteria, it was not taken into account in the investigation. Note that in all studies conducted in the present work, except for the investigation on dielectric electro-active polymers, any test run needs to be related to a reference data set. Both data sets consist of 20 pressure measurements and all need to be related to one Re_b to enable a comparison of the two. Otherwise the reproducibility would limit the identification of small changes in skin-friction between the test cases. Therefore, the ambient values of T, p and, if available, h were minutely documented for every single set of measurement.

4.4 Steadiness of the flow

Differences in reproduced data result not only from the non-repeatability of the flow rate measurement or the measurement of the pressure loss, but also depend on steadiness of the flow rate. If the wind tunnel does not produce exactly the same flow rate over the time of measurement this can have a certain influence. At first sight the steadiness should have no impact on the accuracy, since a deviation in Re_b will cause a corresponding deviation in Δp. It may thus not be possible to exactly reach a certain Re_b but all data should match the curve $c_f = f(Re_b)$. Now if a measurement of a single tab is conducted, the effect may be without impact. In the case of a series of readings to determine a distribution along the channel, the Reynolds number can slightly change with time. In consequence, any single data point of the data corresponds to a slightly different Re_b and plotting the development of the flow along the channel can show considerable scatter. The question arises how long one pressure tab needs to be recorded to avoid the time-dependence. To answer this question, the flow rate in the facility is measured over 10min, which is estimated to be long enough to obtain $c_f \neq f(t)$. During the measurement the ambient parameters are reviewed to be exactly constant. The 10min-interval is then split into defined time periods to compare the mean value of each section to the 10min-average. Exemplarily for a time step of 1min, we get 10 intervals and identify the one with the largest deviation. As the steadiness is depending on Re_b a measurement for different flow rates is required to represent the full Re_b-range. In Figure 4.14 the results for the turbulent flow conditions of three minutes intervals is shown (13% tripping, see Chap. 5). For $Re_b > 8000$ the influence is extremely small, but it slightly grows for lower Re_b. The deviations is still smaller 0.1% even for $Re_b = 2700$, which corresponds to the transitional regime. Thus a recording time of 3min was found to be adequate to exclude time dependence.

There is another aspect of time-dependent fluctuations of the flow left out by the consideration so far: the interaction of the tunnel and the test section. If the friction value in the test section is not constant, the ventilator has to deal with varying pressure losses and the wind tunnel control needs

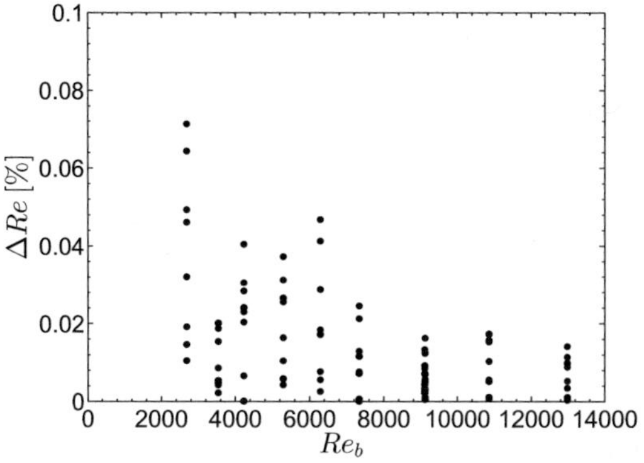

Figure 4.14: Deviation of 3min minutes sequences of a 10min measurement of the Reynolds number.

to react to keep the flow rate constant. For laminar as well as for turbulent flow this effect is negligible, but for transitional flow conditions the wind tunnel control becomes a key parameter. In the present study very good results have been achieved by setting the tunnel control to simply keep the rotation rate constant. The described effect should not be confused with the data scatter in the transitional region. The transitional regime naturally contains fluctuations of the pressure losses for the same flow rate, which can not be filtered out by increasing the integration time and even is not an result of the tunnel control. This random unsteadiness is part of the transitional flow behavior and cannot be considered a measurement error.

5 Reference measurement

The present Chapter considers the measurement results obtained with a smooth surface in the test section. The main goal of the Chapter is to generate a reference data set for the smooth surface, which is to be compared with literature data. Two fundamentally different flow conditions are investigated:

- naturally developing channel flow
- artificially generated turbulent channel flow.

The naturally developing flow is the important reference for laminar and transitional flow control. Throughout the Chapter data with focus on the transitional range of Re is presented. These data of the flow will be exploited for the reference case in Chapter 7.

The artificial generation of turbulence aims at producing a fully developed flow at rather small Reynolds numbers. In such a range of Reynolds numbers, the flow would naturally be laminar or transitional. Considering surface morphology that has shown to be effective for reducing skin friction in turbulent air flows, the physical dimensions of the surface structure were found to be quite small and to decrease with increasing Re. Therefore it is particularly beneficial to work at low Reynolds numbers to avoid a Reynolds number range were fabrication tolerances would influence the results. This demands for initiating transition to turbulence at low Re. Due to the care, which has been taken to achieve undisturbed flow conditions for laminar investigations, artificial tripping is essential to enable transition to turbulence at low Re. On the one hand, the flow is forced to be fully turbulent at low flow velocities and on the other hand the development length is shortened.

For both approaches hot-wire-instrumentation is used to determine the flow state at the channel outlet and is complemented with the results observed with the pressure tap measurement along the channel. The latter provides information about the flow development and the skin-friction. Comparison with literature data and theoretical predictions introduced in Chapter 2 is provided.

Furthermore, three of the initial assumptions require experimental examination:

- the deviation from an ideal two-dimensional flow
- the development length in the turbulent case
- the verification of a hydraulically smooth surface.

The development length is investigated and compared with literature data for turbulent and laminar flow.

The reference measurement will indicate whether the hydraulic smooth condition is justified with $Ra = 0.4\mu\text{m}$. Good agreement with the solution for smooth plates is expected since surface roughness is 10 times smaller than the strict criterion for turbulent flows obtained from Grüneberger's investigation (Grüneberger & Hage, 2011).

The first two aforementioned aspects are clarified through the presentation of the reference data. The quantification of the spanwise distribution of velocity and skin friction demands for an introduction of a special measurement technique, the Preston tube, and is separately presented at the end of this Chapter.

5.1 Naturally developing flow

The measurements for laminar and transitional flows are presented in more detail by Schickel (2012). Only a short summary will be given in this section introducing the main aspects of the solid reference case. Since in the meantime a calibration of the 50mm-nozzle, which was used for the flow rate measurement by Schickel (2012), enabled a more accurate computation of the flow rate, the data were replotted.

First we study the turbulence level Tu in the channel in dependence of Re_b using a hot-wire installed approximately 70mm upstream of the channel outlet. All hot-wire measurements were conducted with a Dantec Streamline system. Utilizing the formula

$$Tu = \frac{\sqrt{\overline{u^2}}}{\overline{U}} \cong \frac{4\overline{E}}{\overline{E}^2 - \overline{E_0}^2}\sqrt{\overline{e^2}}, \tag{5.1}$$

which is derived in Appendix A.2, allows to directly compute Tu from the voltage output E of the anemometer. In the formula $\overline{E}$ is the time average of the voltage, $\overline{E_0}$ the voltage output with resting fluid and $\sqrt{\overline{e^2}}$ the RMS-value. The major advantage of this procedure is the avoidance of time-consuming hot-wire calibrations.

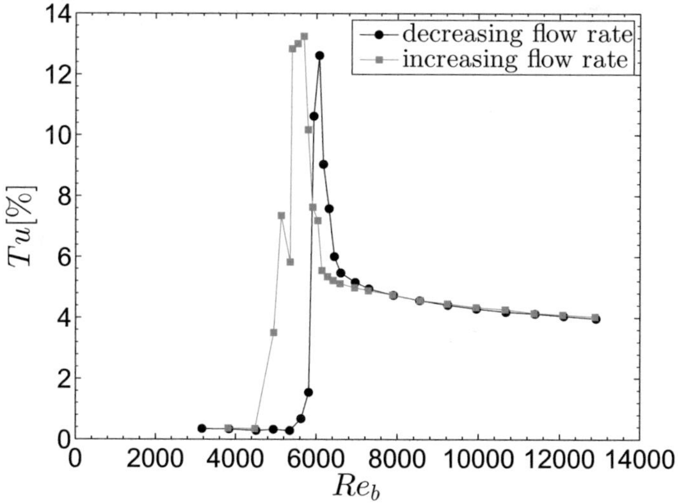

Figure 5.1: Turbulence level in naturally developing channel flow in dependence of Re_b for increasing and decreasing flow rate.

Figure 5.1 shows the turbulence level in dependence of the Reynolds number for increasing and decreasing Re_b. The flow rate has been changed in very small steps to exclude affecting the flow state by acceleration or variation of the flow rate. As can be observed at low Re_b the turbulence level is small, indicating laminar flow. It abruptly increases when reaching the transitional flow state at approximately $Re_b = 6000$ for increasing flow rate. In the fully turbulent state the turbulence level asymptotically approaches to $Tu \approx 4\%$ for $Re_b > 7000$ which agrees with the results of Durst *et al.*

(1998). For decreasing flow rate the transition region is slightly shifted to smaller Re_b. This behavior is typical for naturally developing flow and was observed by Jovanović *et al.* (2011) in the same range of Re.

If we regard the flow development in the channel for different Re_b (Figure 5.2 (a-f)) different streamwise developments of the flow are observed. For low Reynolds numbers (a) and (b) a rather constant slope is reached at the channel outlet. At higher Re_b the development length increases and exceeds the length of the tunnel. The onset of transition in the downstream part of the channel at $Re_b = 6190$ (c) can clearly be noticed from the pressure distribution. The onset of transition results in an inflection point in streamwise distribution of Δp, which moves upstream with increasing Re_b (d-f). At $Re = 12800$ (f) transition to turbulence is still observed over a considerable length of the channel.

In Figure 5.3 the pressure loss of the experimental data for the complete length of the test section are compared with the theoretical prediction for laminar developing flow by Schlichting (1934), which has been introduced in Chapter 2.2.2. Since measurements by Durst *et al.* (2005) proved that the development length is slightly longer than Schlichting's prediction (see Fig. 5.4), the result from the theory has been modified for a 5% longer development length in Figure 5.3. Although the theory is based on a two-dimensional theory, the agreement is rather good. If one accepts Schlichting's approach, assuming a centerline velocity deviating 2% from the value of the parabolic profile to be negligible, we can compute the development length. The result is plotted in Figure 5.4. Additionally, for four marked points in Figure 5.4, the corresponding pressure loss measurement are shown in Figure 5.5(a-d). In these plots the analytic solution $c_f = 12/Re_b$ is included by straight lines and is compared with the slope of the pressure drop at the channel outlet: The length of the channel is not sufficient to fully develop the flow for $Re_b > 3600$.

Complementary, Figure 5.6 shows the mean value of the friction coefficient for the complete test section (a) and the downstream segment (b) of test section. The transitional region can be clearly identified by a large increase of c_f in both plots. Since (b) shows larger values of c_f for the complete turbulent range, the persistence of a distinct fraction of laminar flow in the channel is confirmed even for highest Re_b. The comparison to Dean's law in the downstream section in (b) indicates that for the highest Re_b a fully developed state is achieved in the downstream segment. In the laminar regime is it clearly visible that the measurement data departs from the

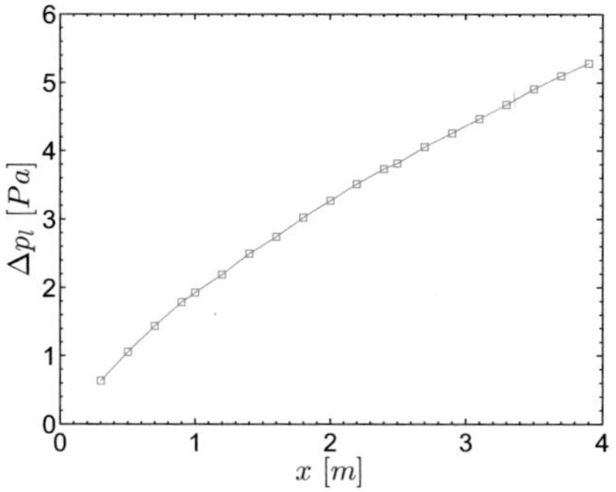

(a) $Re_b = 4250$

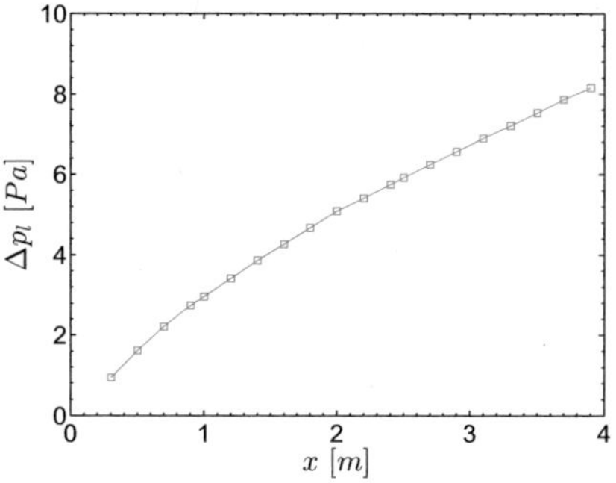

(b) $Re_b = 5820$

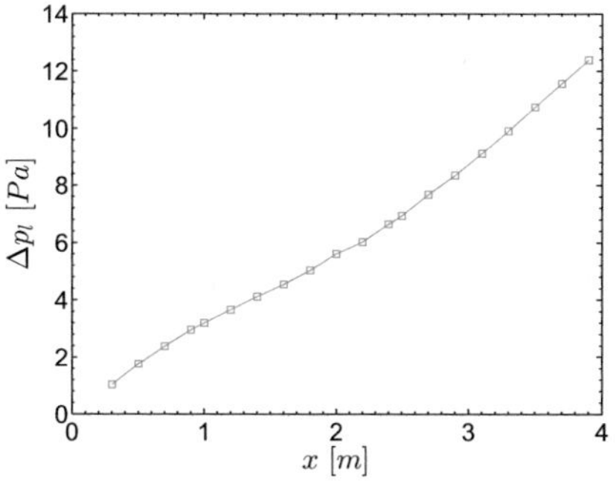

(c) $Re_b = 6190$

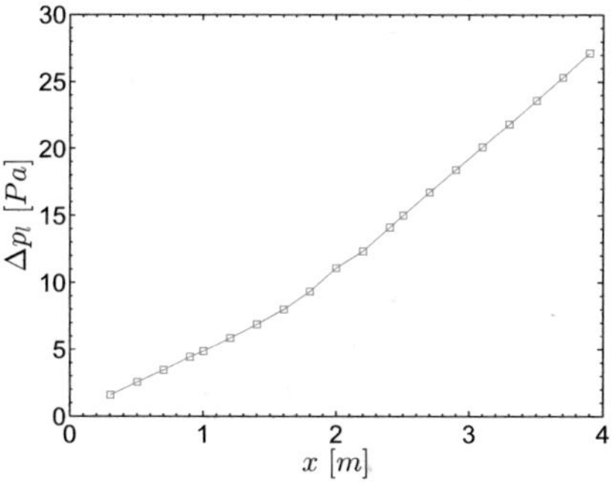

(d) $Re_b = 8180$

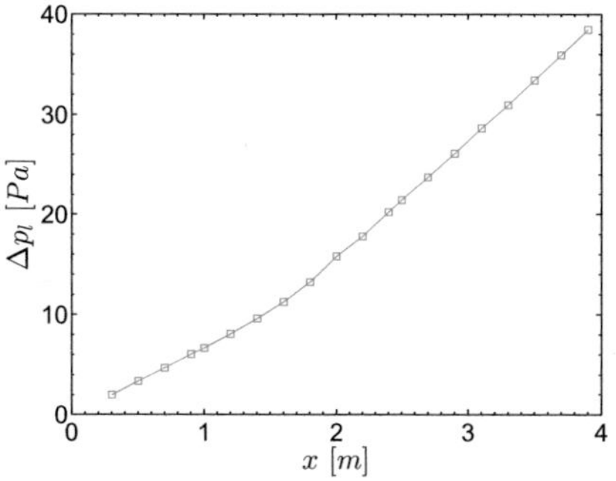

(e) $Re_b = 10000$

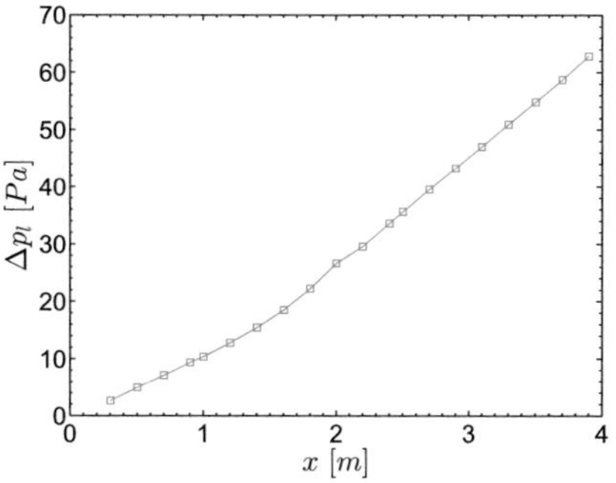

(f) $Re_b = 12800$

Figure 5.2: Streamwise distribution of the static pressure in the test section for different Re_b.

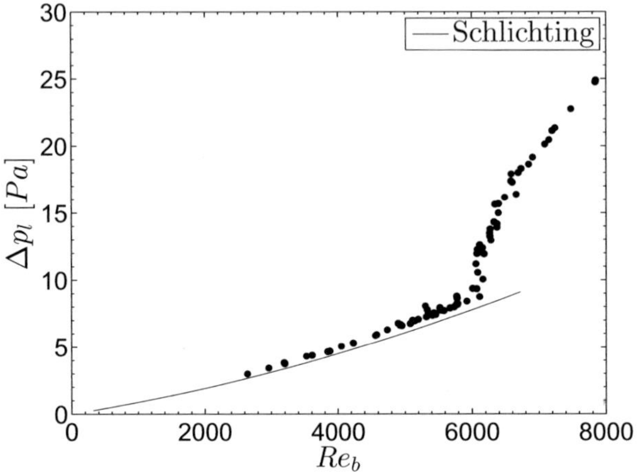

Figure 5.3: Comparison of a theoretical channel entrance (Schlichting (1934)) flow with the measured pressure loss distribution.

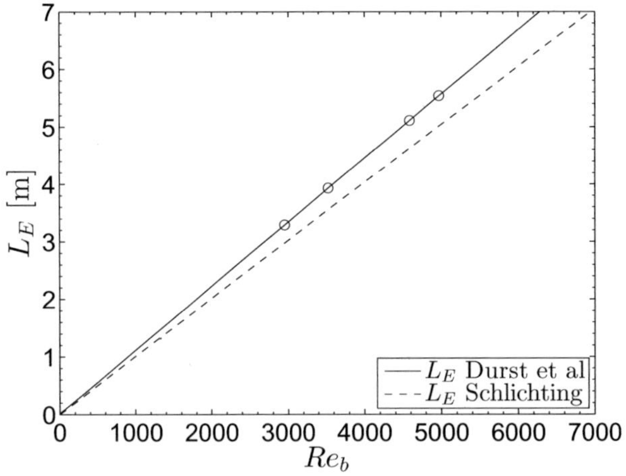

Figure 5.4: Development length of a laminar channel flow calculated for the dimensions of the present channel according to Schlichting (1934) and Durst *et al.* (2005). Circles denote measurement points depicted in Figure 5.5.

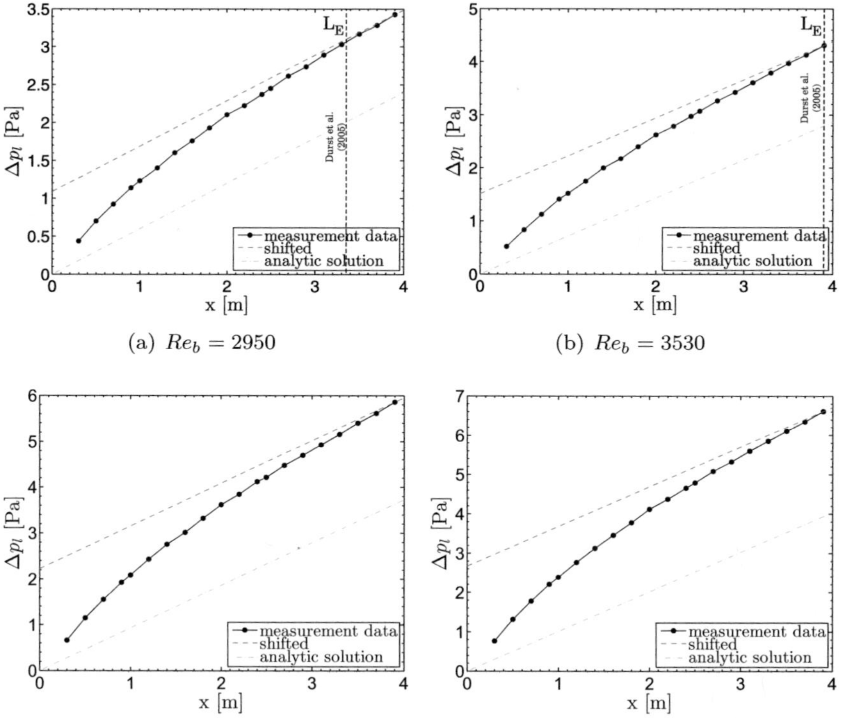

(a) $Re_b = 2950$ (b) $Re_b = 3530$

(c) $Re_b = 4590$ (d) $Re_b = 4970$

Figure 5.5: Streamwise development of laminar flow at different Re_b. Dashed lines mark the development length L_E according to Durst *et al.* (2005) in (a) and (b).

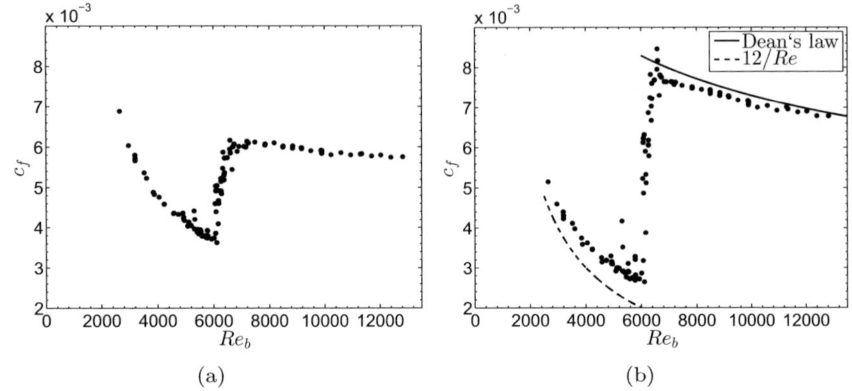

Figure 5.6: Friction coefficient calculated for the total length (a) and the downstream segment (b)

analytical solution with increasing Re_b, which is not surprising since the running length is not sufficient for a full development as seen before.

Note that in all diagrams of the naturally developing flow the pressure tap at 2m is obviously working imperfectly. Since the deviation is small and reproducible, the obtained information is still valuable for the investigations.

Finally, we inspect the data of the natural flow development in Figures 5.2 and 5.5 with respect to disturbances at the intersections of the channel plates at 0.95m and 2.45m. There is no indication that the transition is triggered by imperfections at the intersections for any value of Re.

5.2 Artificially generated turbulent flow

In comparison with the previously presented study of naturally developing flow, the set-up has been modified in two significant details:

- Two Baratron sensors measure the pressure along the test section, which almost saves half of the time. The pressure taps at the first and middle position are read by both sensors simultaneously to detect potential deviations between the devices.
- The temperature measurement has been improved and the humidity is considered for the studies of riblets and dielectric electro-active polymer actuators.

Different tripping has been tested to achieve turbulent flow for lowest Re. The turbulence level has been measured for 12%, 13% and 16% blockage of the cross sectional area. Each tripping device was made from a 0.35mm thin brass section and installed at the top and bottom wall directly after the contraction in the tunnel entrance.

Figures 5.7 and 5.8 show the corresponding results for the turbulence level measured with a hot-wire at the end of the channel and the friction coefficient for the last section of the channel ($x > 100H$), where the development length can be assumed sufficient for turbulent conditions (c.f. Chap. 2.2.2).

Figure 5.7 (a) shows the turbulence level for the 13% tripping. Transition is observed at approximately $Re_b = 2300$. The hot-wire measurements indicate that a fully turbulent flow at the outlet develops at $Re_b = 3500$. In Figure 5.7 (b) the friction coefficient for the last segment of the channel for the same Reynolds number range is illustrated. The comparison to Dean's law (line) implies that a fully developed state is reached for $Re > 3500-3800$ on the last segment, which corresponds to higher flow rate than desired. With 12% blockage (not shown) the result is similar; both the transition as well as the agreement to Dean is shifted to even higher Reynolds number ($\Delta Re_b = 200$). Thus the tripping was increased to 16%, which leads to turbulent flow at even smallest Re_b (Fig. 5.8(a)). This is confirmed by the good agreement to Dean on the last segment (b). Yet, a disadvantage at the high tripping level has been observed. The large tripping device does not only affect the boundary layers of the developing flow and triggers transition, it also causes large fluctuations in the static pressure. This can be interpreted as a phenomena similar to the flow downstream an orifice

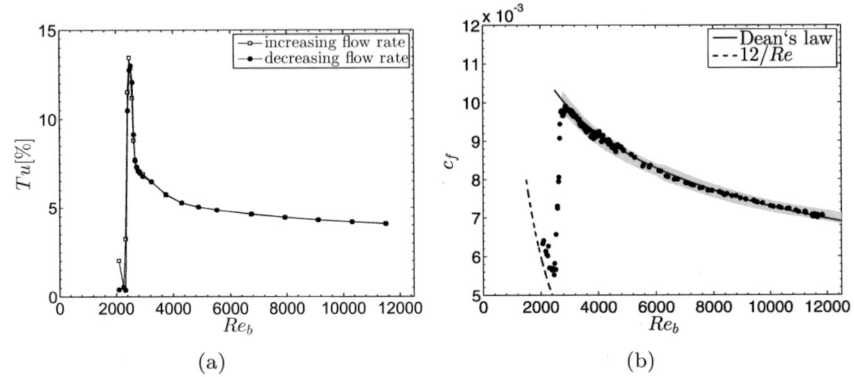

Figure 5.7: Turbulence level at the channel end (a) and friction coefficient over Re_b on the downstream segment (b) for the tripping with 13% blockage. In (b) Dean's correlation is compared with the data and uncertainty intervals deduced in Chapter 4.3.1 are shaded in grey.

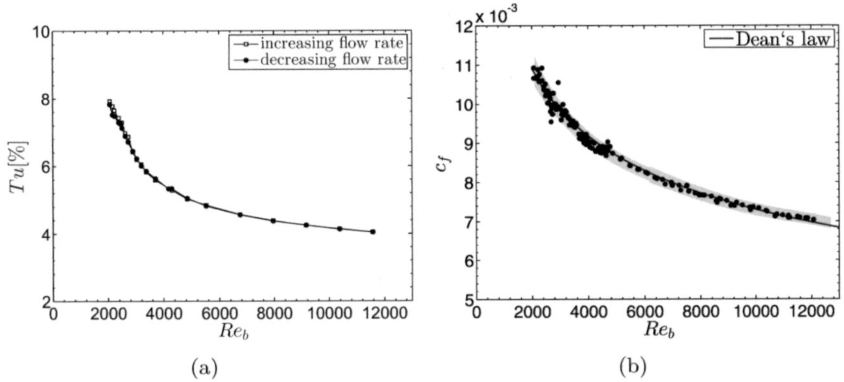

Figure 5.8: Turbulence level at the channel end (a) and friction coefficient over Re_b on the downstream segment (b) for the tripping with 16% blockage. In (b) Dean's correlation is compared with the data and uncertainty intervals are shaded in grey.

disk in a pipe, where a wake region with large vortex structures persists over a long streamwise distance. In Figure 5.10 (b), where the c_f-value for the entrance region (based on the differential pressure at the second tap $x = 0.3$m) is depicted, the resulting large scatter in the data can be observed for small Re_b. In (Fig. 5.8) (b) additionally the uncertainty intervals (grey shaded) are compared with the measurement data. If we consider, that the bias error of c_f was estimated to be larger than 50% of the maximum uncertainty (compare Chap. 4.3.1), the obtained scatter (random error) in case of the 16% tripping significantly exceeds the error calculation for $Re < 4500$. This obviously confirms the "orifice" effect and led to the decision to work with a smaller tripping level.

While for the 13% tripping in Figure 5.7 (b) for turbulent flow the agreement with the experimental literature data is very good, a small deviation from the analytical solution for laminar flow $c_f = 12/Re_b$ is obtained. There are three potential sources why the c_f of the channel is slightly higher than the analytical solution at low Re ($Re_b < 2500$) in Figure 5.7(b). The first is the development of the flow, where a fully developed flow state is not completely accomplished and can lead to slightly higher Δp_l and consequently c_f. The second and estimated to be the dominant influence is the deviation from a two dimensional flow state. The third reason is the disturbance created by the tripping, which may not have dissipated completely, thus generating some higher friction. However, this influence is probably negligible in the second half of the channel.

The region of increased scatter between $3800 < Re_b < 4400$ results from characteristics of the 50mm-nozzle at low flow rates, which was identified after the campaign and is discussed in Appendix A.1. The increased scatter basically results from transition in the nozzle. The mean value is not affected, while the scatter is approximately twice. Therefore, the orifice meter was used for later measurements with 13% tripping, as in the case of the riblet and the dielectric elastomer actuator studies. Nevertheless, the results for the first generation of the instrumentation show less scatter than most of the available literature data, but the benefit of the recent evolution is clearly visible in Figure 5.11.

The 13% blockage is considered to be the best compromise between shifting the fully developed turbulent flow to lower Re and introducing large scale disturbances. Therefore, it is chosen for tests in the studies presented in Chapters 6 to 8. In the following the discussion of the flow development will focus on this configuration.

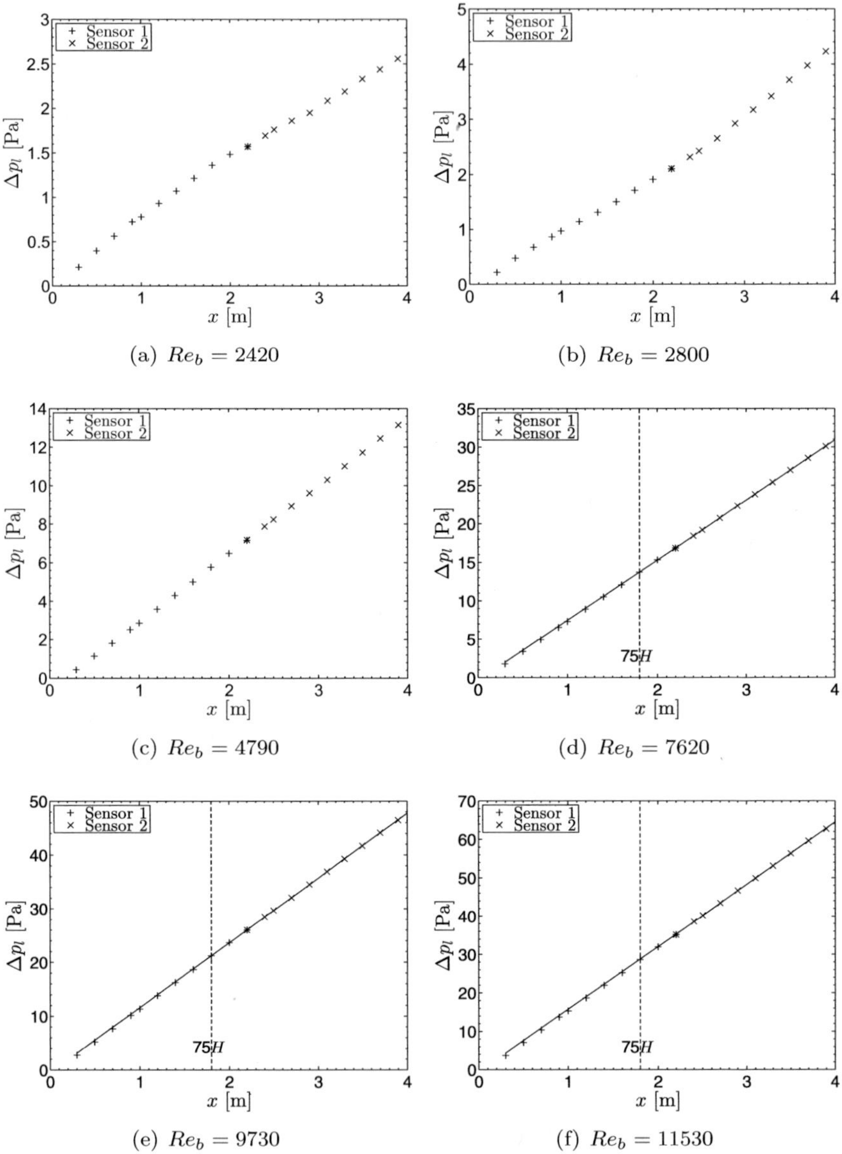

(a) $Re_b = 2420$ (b) $Re_b = 2800$

(c) $Re_b = 4790$ (d) $Re_b = 7620$

(e) $Re_b = 9730$ (f) $Re_b = 11530$

Figure 5.9: Streamwise distribution of the static pressure for different Re_b (13% tripping). Position 11 is measured by both sensors simultaneously. For fully turbulent flow in (d-f), lines show least-square-fits though the last 10 taps. The dashed line marks L_E.

The reference measurement with 13% tripping was repeatedly reproduced over two years of measurement. While the data presented in the following paragraph were measured without considering the humidity, the data for higher Re_b, which are shown later, correspond to the final measurement instrumentation. In Figure 5.9 (a-f) the development of the flow with increasing Re_b is depicted at four characteristic points. In (a) disturbances are dampened and the flow remains laminar over almost the entire length at the low $Re_b = 2420$. Although transition is induced in (b) the flow remains in an intermittent flow state changing between laminar and turbulent for most of the channel length. In (c) the Reynolds number is high enough to reach a fully turbulent state at the downstream segment. With increasing Re_b the length of the transitional region in the first section of the tunnel reduces. It cannot be observed anymore for $Re_b > 6000$ (d-f). Interestingly, a transitional region close to the channel entrance is also visible in the case for 16% tripping, as depicted in Figure 5.10 (a) and (b). For a smooth surface a steeper gradient in the downstream part of the tunnel results from laminar or transitional flow (a). In (b) the c_f value based on the measurement at $x = 0.3$m is presented. Laminar-turbulent transition is observed between $Re_b = 3000$ and $Re_b = 7000$, which also becomes apparent through the large data scatter in this region.

The development with 13% tripping is further studied in Figure 5.9(d-f). The lines are linear least-square-fits through the last seven measurement points of the test section. They indicate a completed development for running lengths of about 1.8-2m corresponding to approximately $70 - 80H$. This result is found to be independent of Re_b for fully turbulent flow and indicates longer development length than found in the literature. The very high accuracy of the pressure measurement allows direct identification of smallest differences in the flow development, as changes in the slope. For changes less than 0.3% the present set-up reaches its limit. If a deviation in the slope of about 0.5% is accepted $L_E = 60H$, which was suggested by Durst *et al.* (1998), can be confirmed. Note that taps at position 2m were repaired successfully.

Finally, Figure 5.11 is shown to examine the hydraulic smoothness condition, since highest Reynolds numbers are more sensitive to surface roughness. Following Colebrooks diagram (see Fig. 2.5) roughness is indicated, if the value of c_f approaches a horizontal line with increasing Re. On the contrary the distribution of c_f very closely follows Dean's prediction for a smooth channel. As a result, the reference surface is considered to be hydraulically smooth. In Figure 5.11 the last pressure tap is not taken into account for

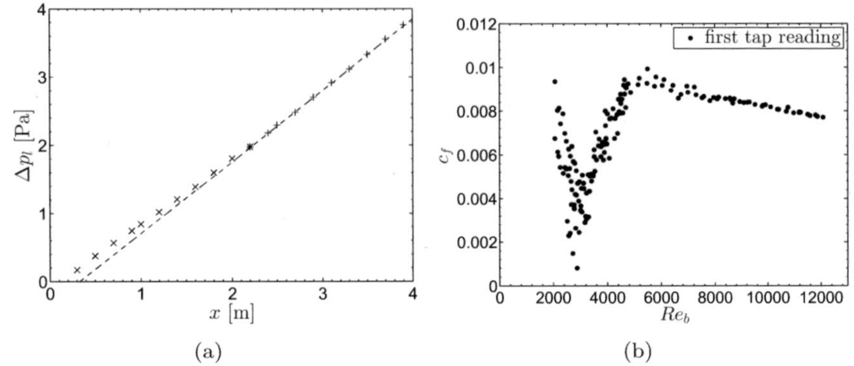

Figure 5.10: Streamwise distribution of the static pressure for different $Re_b = 2440$ (a) and c_f based on measurement of the second pressure tap at $x = 0.3$m with 16% tripping.

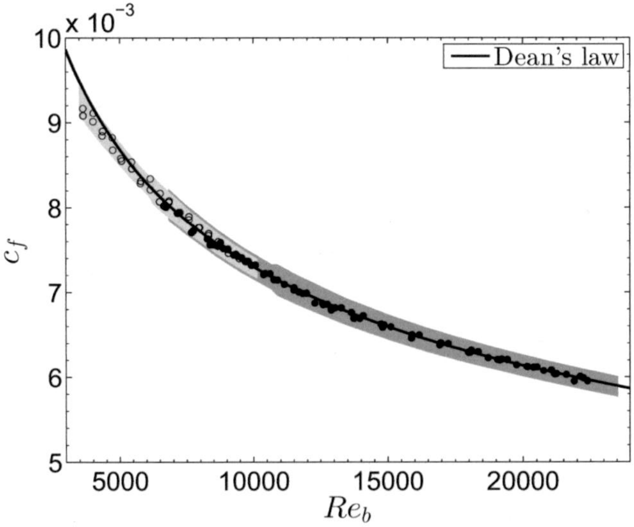

Figure 5.11: c_f over Re_b for the downstream section for final development of the facility. Open symbols indicate measurement with the orifice meter, filled symbols with nozzle. Grey areas show maximum error intervals for each instrument.

the calculation of c_f since a slight deviation of the measurement of this tap at high Re_b was found. The deviation is reproducible and caused by the vicinity to the channel outlet ($\Delta x = 50$mm).

Furthermore, the maximum scatter remains well below the estimated maximum uncertainty, especially for $Re_b > 8000$. This is in perfect agreement with the uncertainty estimation, which has predicted a very good reproducibility (compare Chap. 4.3.1).

5.3 Impact of the laboratory environment on the flow development

The transition measurements were very sensitive to noise and vibrations. If combustion engines or compressors were operated in neighboring laboratories, significant shifts of the transition region towards smaller Re were observed. Not the noise itself, but the vibration transferred by the groundwork were deemed to be more critical. Figure 5.12 compares a measurement for quiet and slightly noisy conditions both at $Re_b = 5600$. While the flow remains complete laminar in (a), transition is triggered in (b) resulting significantly higher pressure loss in the downstream part. It can be clearly observed that the standard deviation of the pressure increases considerably with the onset of transition. Not only the downstream part, but the entire length is affected by the intermittent state of the flow, leading to fluctuations in the pressure drop. In consequence, the onset of transition has been checked before every measurement and monitored during the measurement time to ensure reproducible conditions. If the transition was triggered by disturbances the measurements were stopped and repeated under calm conditions. The frequent disturbances generated by other laboratory tests required that the investigations were mainly conducted at night time. Figure 5.13 confirms the reproducibility under quiet conditions depicting exemplarily the onset of transition for three measurements of the turbulence level from different dates.

Although the studies with artificial tripping were initially expected not to be affected by environmental noise, a slight influence was obtained for low Re. When experiments were conducted close to or in the transitional Re-range, where a distinct length of development from laminar to turbulent flow was observed (compare 5.9(b)), it turned out that the laboratory environment was able to influence the pressure drop in the channel even for the tripped flow.

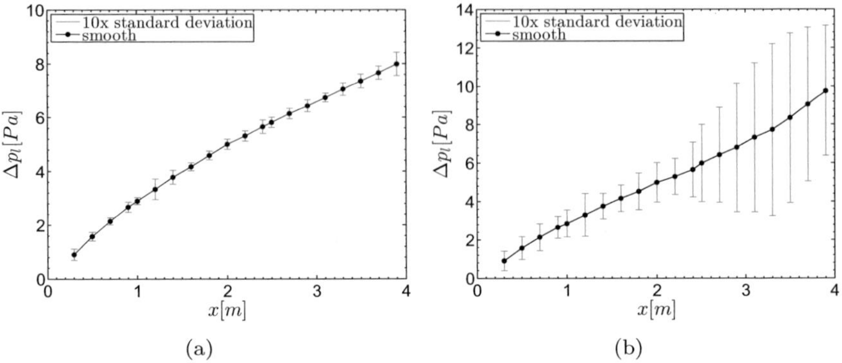

Figure 5.12: Distribution of the pressure loss for quiet conditions (a) and for small disturbances (b).

Figure 5.14 shows the turbulence level in dependence of Re_b, obtained at two distinct days, with 13% tripping. The onset and the maximum peak of the turbulence level differ in Reynolds number ($\Delta Re_b = 100$), for increasing as well as for decreasing flow rate. One day, March 10 2013, was a working day and the activities from the neighboring laboratories, influenced the transition at low Re. It has been found, that noise can have a small influence on the pressure drop, mainly on the upstream section by an amount of approximately 2-3%. For $Re_b > 4000$ the influence of noise becomes insignificant. To guarantee a fully turbulent over the complete channel length independent of the surrounding conditions, $Re_b \geq 4800$ is required. Thus, data for low Re were very carefully examined to identify any disturbance that might lead to a misinterpretation of the data, as will be shown in following Chapters.

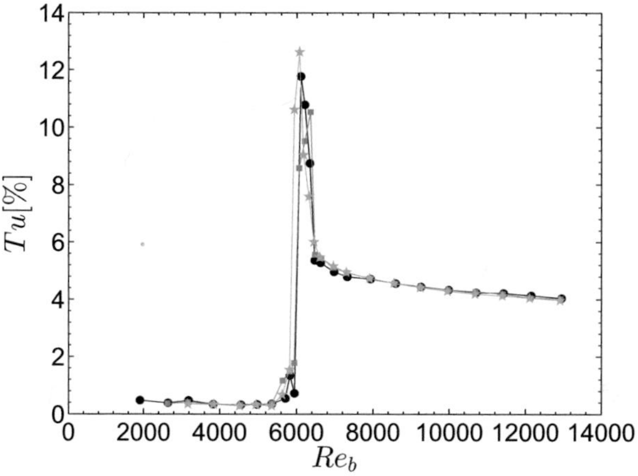

Figure 5.13: Reproducibility of measurement for undisturbed conditions for increasing Re_b from three different measurement days.

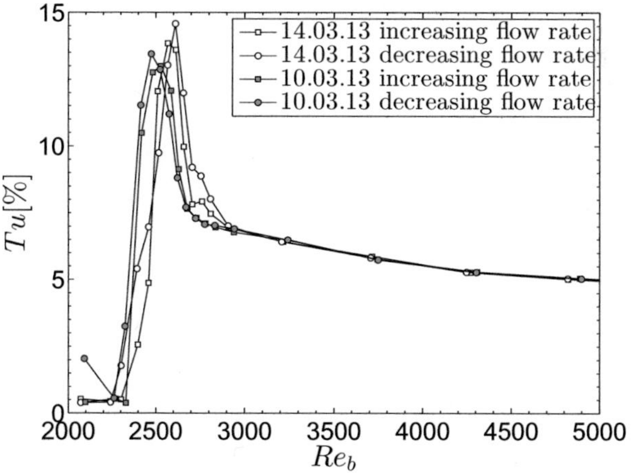

Figure 5.14: Reproducibility of measurement for 13% tripping for increasing and decreasing Re_b from two different measurement days.

5.4 Two-dimensionality of the flow

In Chapter 2 the necessity to justify the assumption of two-dimensional flow for a real, high-aspect ratio duct flow became evident. In order to analyse the deviation from the ideal flow assumption, the wall shear stress distribution in spanwise direction needs to be detected for the bottom wall of the channel. This requires an appropriate experimental technique, which contrary to the available pressure drop measurement can resolve the distribution of τ_{wall} in spanwise direction. A method very simple to implement is the Preston tube. The standard type is a Pitot tube of small diameter positioned directly at the wall. The Preston tube uses the dynamic pressure in near wall region as a measure of the wall shear stress. Being a single-point method, it allows to resolve the shear stress distribution by spanwise traversing the probe. Advantageously the required high accuracy pressure transducer is already part of the equipment.

As will be explained in the following the use of a universal function for the Preston tube measurement is limited to the fully turbulent flow, otherwise a calibration is required. In consequence the experimental examination of three-dimensional effects does not include laminar and transitional flow.

The following Section investigates the spanwise distribution of specific flow quantities in turbulent flow conditions to identify:

- the centerline velocity via Pitot tube
- the wall shear stress via Preston tube.

The measurement data shown in the previous Section confirmed the high measurement accuracy, which was calculated in Chapter 4. However, this is the particular measurement of an $AS = 12$ channel, which for comparison to the literature has been calculated on the two-dimensional theory according to Dean (1978). Three-dimensional effects caused by the presence of the sidewalls have not been considered. As soon as a technique for drag reduction is applied, the deviation from a uniform distribution of the skin friction influences the comparison of the skin-friction with the smooth reference $\tau_w/\tau_{w,0}$. Additionally it must be considered that the application on the side walls is not possible. While for very large aspect ratios the difference to a two-dimensional flow is negligible, in the range of commonly used moderate aspect ratios of 8-20 the question arises to which degree the results for $\tau_w/\tau_{w,0}$ are influenced.

The goal is to measure the spanwise distribution of τ_w and to utilize the knowledge to correct for the 3D-effects to provide better comparisons with ideal 2D channel flow which are typically obtained from analytical considerations or direct numerical simulations.

5.4.1 Pitot tube measurement

Since a Pitot tube needs a static reference pressure, it was positioned precisely at the streamwise position of the second to last pressure tab to obtain the corresponding static pressure. In Chapter 2 it was deduced that in turbulent channel flows, in contrast to laminar flow, the static pressure in x_2-direction deviates from that on the wall. A similar deviation can also be expected for the spanwise centerline (cf. Eq. 2.27). Therefore, the spanwise distribution of the static pressure was checked to ensure that the velocity with the pitot tube is not influenced by static pressure variations. The variation of the static pressure has been measured at $x = 3700$mm with a static tube traversed in spanwise direction on the centerline. Difference pressure was obtained by using the reference from the corresponding wall tap. The Baratron 698A has been used with a 1.3Pa range (accuracy 0.12% R) recording each position for 3min.

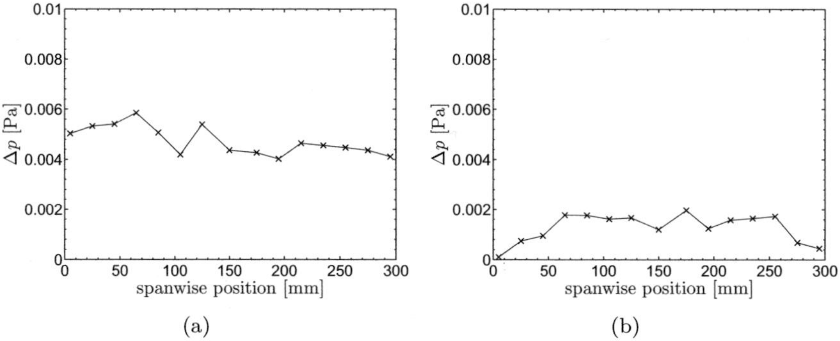

Figure 5.15: Distribution of the difference between static pressure at the wall tab and the static probe for $Re_b = 8770$ (a) and $Re_b = 15100$ (b) along the spanwise centerline.

The results in Figure 5.15 (a) and (b) indicate diminutive deviation from the wall tab reading. The deviation of a few Millipascal is within the

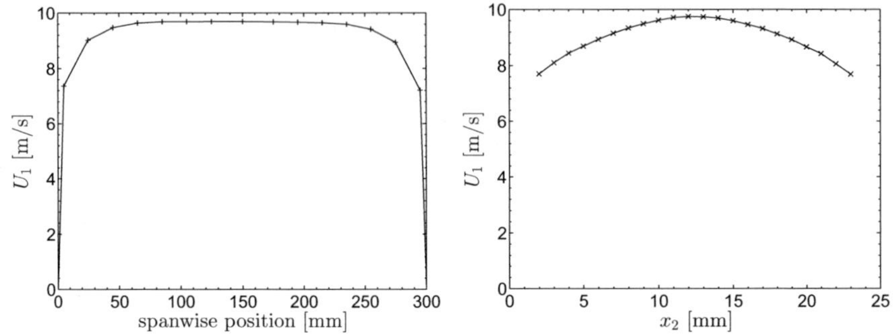

(a) (b)

Figure 5.16: Velocity distribution on the spanwise (a) and vertical centerline (b) from Pitot tube measurement with the wall tab as reference pressure at $Re_b = 13650$.

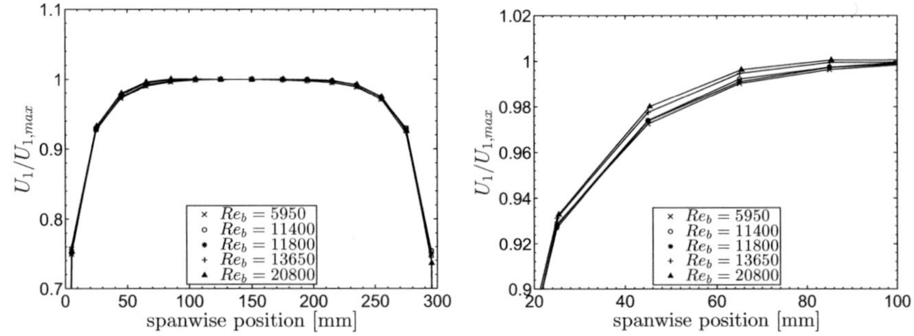

(a) (b)

Figure 5.17: Spanwise measurement with Pitot tube for different Re_b normalized with $U_{1,max}$. The complete channel width is shown in (a), while (b) depicts a detailed view on the region near the sidewalls.

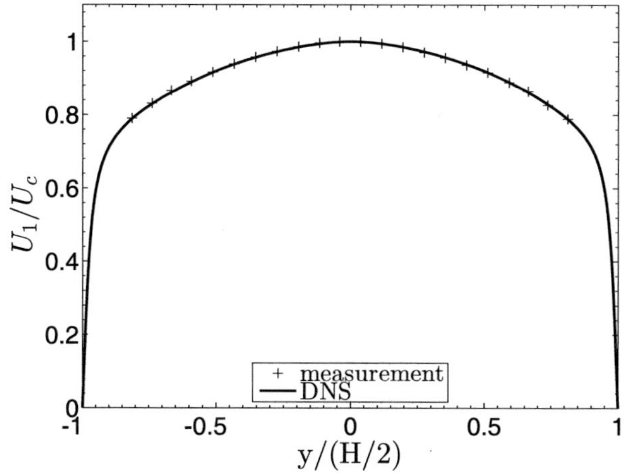

Figure 5.18: Comparison of DNS and wall-normal measurement with Pitot-tube for $Re_b = 13650$ normalized with U_c.

range of measurement uncertainty (the uncertainty of metering zero is higher). The measurement result with the static tube for $Re_b = 8870$ (a) and $Re_b = 15100$ (b) correspond to dynamic pressures of $p_{dyn} \approx 30$Pa and $p_{dyn} \approx 60$Pa. Thus the static pressure on the spanwise centerline is confirmed to be virtually constant and the pressure tab is used as static reference.

For the measurement along the centerline in x_2 and x_3-direction a 1.3mm Pitot tube has been used; again measurements are taken at $L = 3700$mm. In combination with the Baratron sensor an accuracy significantly better than 1% can be estimated. Figure 5.16 depicts the result for $Re_b = 13650$ on spanwise and vertical centerline. The vertical data begin and end in a distance of 2mm from the wall to minimize the displacement effect due to the presence of the Pitot tube.

In Figure 5.17 (a) the spanwise distribution is shown for a selection of Re_b. The velocity is normed with $U_{1,max}$, which has been computed from the three recordings at the channel center. In general the distribution of U_1 is almost identical for all considered Re. A detailed view in Figure 5.17 (b) identifies a slight change toward the ideal, two-dimensional distribution with increasing Re_b, which is in agreement to the fundamentals presented in Chapter 2. In general approximately 70% of the channel width show a nearly constant spanwise velocity distribution. Approaching the sidewalls the centerline velocity decreases to $0.75 \cdot U_{1,max}$ at a 5mm distance from the corners.

Since the measurement data deviates by 1% from the Reynolds number of the comparative DNS of periodic channel flow by Moser *et al.* (1999), a high accuracy DNS (Luchini & Quadrio, 2006) for exactly $Re_b = 13650$ was especially computed by Gatti (2014a) to compare the vertical measurement data with a two-dimensional flow field. In Figure 5.18 all data are normalized with U_c and the measurement data are corrected with the procedure proposed by McKeon *et al.* (2003) and Bailey *et al.* (2013) for a Pitot tube measurement in the near wall region. The agreement of the available measurement section is very good. A fully developed flow state is confirmed, which exhibits the same shape as the two-dimensional simulation. This result enables to compare the flat duct with the ideal channel flow in Section 5.4.3.

5.4.2 Preston tube measurement

The Preston tube requires a calibration for the relation between the dynamic pressure at the wall and the wall shear stress. As shown in Chapter 2.1.2, in a fully developed turbulent flow the velocity distribution in the near wall region follows universal laws. For pipe flows, channel flows and boundary layer flows, the velocity in the viscous sublayer can be described by a linear distribution, whereas the velocity in the adjacent region follows a logarithmic distribution. Preston's idea was to find a universal calibration curve on base of the logarithmic law to use this for general measurement purpose. From dimensional analysis the functional relation

$$\left(\frac{U_1}{u_\tau}\right) = B + C + A \cdot log\left(\frac{u_\tau x_2}{\nu}\right) \tag{5.2}$$

can be derived. With B, C and A being constants (Ferriss, 1965). For a the relation between the pressure difference of a wall tap and a Pitot tube located in the logarithmic layer Preston (1954) derived the relation

$$\left(\frac{\Delta p d^2}{4\rho\nu^2}\right) = f_1\left(\frac{\tau_w d^2}{4\rho\nu^2}\right). \tag{5.3}$$

The function f_1 in equation (5.3) was calibrated with experimental data. Preston (1954) demonstrated the validity of his concept and suggested the following calibration function:

$$y^\star = 0.875x^\star - 1.396 \tag{5.4}$$

$$\text{with} \quad y^\star = \log\left(\frac{\Delta p d^2}{4\rho\nu^2}\right) \tag{5.5}$$

$$\text{and} \quad x^\star = \log\left(\frac{\tau_w d^2}{4\rho\nu^2}\right) \tag{5.6}$$

Besides the diameter of the tube d the function depends on the fluid properties. Obviously small uncertainties in viscosity or density have large influences on the result. Ferriss (1965) and Head & Rechenberg (1962) verified the procedure and found slightly diverging constants.

A Preston tube would ideally be located in the log-law region of turbulent wall-bounded flow. This is achieved for large Reynolds numbers, but for low *Re* the tube orifice contains part of the buffer layer region. To enable a wide range of applicability of the function, this effect needs to be taken into account. Bradshaw *et al.* (1961) discussed that a relation of the form of equation (5.3) even exists for the viscous sublayer. Thus Bechert (1995) and Patel (1965) refined the calibration based on more complex formulas. While Bechert developed a single formula for a far range of measurement, Patel defined three ranges of $y^\star$ to include a wide measurement range. Following the advice from Sutardi & Ching (2001), where both approaches have been compared, the formula by Patel has been used for the present computations. For the range of the present investigation Patel's formula writes:

$$y^\star = 0.8287 - 0.1381x^\star + 0.1437x^{\star 2} - 0.0060x^{\star 3} \tag{5.7}$$

A Preston tube of $d = 1.3$mm with a simple circular shape has been used. Since the tube needed to reach 250mm into the channel and still guarantee adequate mechanical stability of the tube, it was custom-built of three tubes of growing diameter with conical intersections (1.3, 2, 3mm). The total length is 310mm. 35mm is the length of the smallest diameter tube at the tip and 250mm the length of the tube with the largest of 3mm outer diameter.

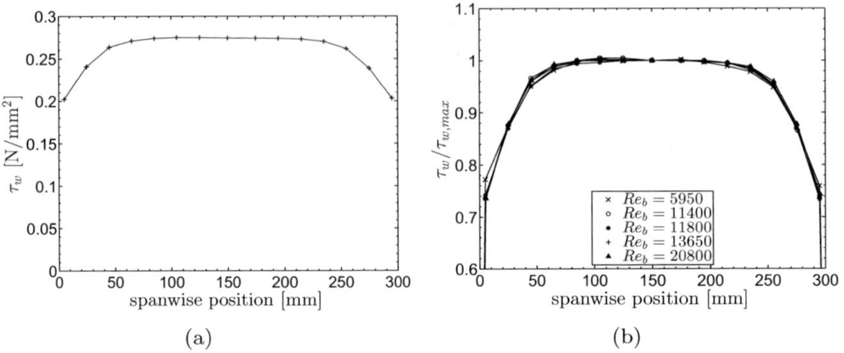

Figure 5.19: Spanwise distribution of the wall shear stress for $Re_b = 13650$ (a) and normalized with $\tau_{w,max}$ for different Re_b (b) from Preston tube measurement with the wall tab as reference pressure.

In Figures 5.19 (a) and (b) the results, based on Patel's function, of the spanwise wall shear stress measurements are shown. For $Re_b = 13650$ the result is depicted in (a). Obviously the distribution of τ_w is not two-dimensional in the corner region. For 70% of the channel width in the center it is close to a constant value, on another 10% of the surface rather small decrease of τ_w is observed. In the vicinity of the corner at 5mm the deviation reaches 25%. At the corners τ_w declines steeply to zero, which only represents a very small fraction of the channel width. In Figure 5.19 (b) the result normalized with $\tau_{w,max}$ is shown for various Re_b. For increasing Re_b again a slight approaching towards the ideal, two-dimensional distribution is observed (compare Sec. 5.4.1). In general there is only a weak dependence of the wall shear stress distribution on Re_b.

The specification of the measurement accuracy of the Preston tube is not straightforward, since it primarily depends on the quality of the calibration function. To obtain an estimate, the spanwise mean $\tau_{w,b}$ from the Preston tube measurement is compared with the integral value of τ_w based on the pressure tab measurement from the reference study for the later riblet investigation in Table 5.1. The integral value and the mean from Preston tube measurement are in good general agreement, since Bechert (1995) gives an accuracy of better than $\pm 5\%$ for a Preston tube. The average difference of 5-6% compared with the pressure tap measurement does not represent

Table 5.1: Comparison between determination of τ_w from pressure drop and Preston tube measurements.

Re_b	$\tau_{w,b}$ (Patel)	τ_w (Dean)	Δ
[-]	[N/m^2]	[N/m^2]	[%]
5950	0.060	0.065	6.5
11400	0.190	0.197	3.5
11800	0.204	0.214	4.5
13650	0.257	0.275	6.5
20800	0.533	0.584	8.5

the measurement error. It is the result of the friction at the side walls, which is included in calculation from the pressure drop. The determination of τ_w from the pressure tap measurements is based on two-dimensional theory neglecting side wall effects. Thus the τ_w value is systematically overestimated.

5.4.3 Influence of three-dimensional effects on the effectiveness of drag reduction techniques

In the following the contribution of the side walls to the pressure loss is considered. This is of importance as the drag reduction of any of the later presented methods will not affect this area. If this effect is not taken into account the obtained amount of $\tau_w/\tau_{w,0}$ is smaller than present on the controlled area. Further the flow in vicinity to the corners will be slower and the drag reduction mechanism will probably not act with same efficiency as in the rest of the channel. The goal is to give an estimate of the deviation from a ideal two-dimensional flow. On the one hand this enables a better comparison with numerical results and on the other hand to predict the real potential of the techniques.

In Figure 5.20 the estimation of the global distribution of τ_w is shown. On the top and bottom wall the distribution of τ_w from the Preston tube measurement for $Re_b = 13650$ is sketched. On the sidewalls the distribution from top and bottom wall is mirrored on the dashed bisecting line of the corner angle. It is assumed that this distribution of the wall shear stress accounts for the complete pressure drop along the channel. In

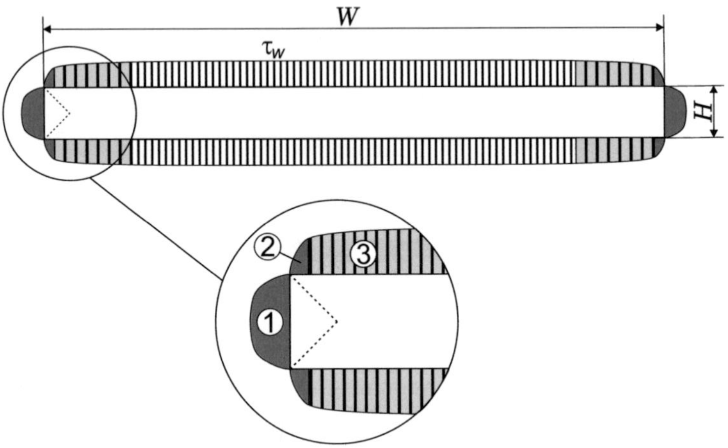

Figure 5.20: Distribution of the skin friction on the surface of the tunnel. Top and bottom wall from measurement, side walls estimated.

the following the assumptions to calculate the decrease of effectiveness of the flow control techniques due to three-dimensional effects are discussed (see also illustration in Fig. 5.20).

1. The sidewalls, where no drag reduction mechanism is applicable, cover 8% of surface. The maximum value of the wall shear stress on the sidewall is 80% of the mean value $\tau_{w,b}$ and reduces to zero at the corners. Therefore the integral value for the sidewalls is half of $\tau_{w,b}$. This accounts for 4% of the total friction. Note that the integral value of τ_w obtained from the pressure tap measurement, which is based on two-dimensional theory neglecting any side wall effects, exceeds the mean value of the Preston tube measurement in the same order of magnitude (compare Table 5.1).

2. It is assumed that any technique does not influence the flow in direct vicinity to corner region of covering plates (compare Fig. 5.20 surface area 2). In a distance to the side walls smaller than 5mm, $\tau_{w,b}$ is very small. The average value less than 40% of $\tau_{w,b}$. On the one hand the influence of any technique will be nearby zero, on the hand this region covers only 3% of the channel surface. The amount is 1.3% of the total skin friction.

3. The decrease of τ_w approaching the side wall region is considered as a local decrease of Re_b. The effectiveness of a technique for manipulating friction drag is a function of Re. None of the investigated drag reduction method accounts specifically to local variations and the decrease of τ_w approximating the to side walls will reduce effectiveness. Approximately 20% of the wetted surface on top and bottom walls exhibit a value $0.7 \cdot \tau_{w,max} < \tau_w < \tau_{w,b}$. In average this is estimated to reduce the effectiveness of the method by 10% (concluded from riblet effectiveness with deviation from optimal Reynolds number, c.f. Chapter 6). This amounts for 1.6% of total skin friction.

In total the underestimation is approximately 7%. If we define

$$\Delta\tau_w = (\tau_w - \tau_{w,0})/\tau_{w,0}, \tag{5.8}$$

the correction writes $\Delta\tau_{w,2D} = 1.07 \cdot \Delta\tau_{w,ch}$ for a mechanism for skin friction manipulation implemented on complete top and bottom walls. For example, applying a passive flow control structure on the entire top and bottom surface leading to a pressure drop decrease of 5%, would correspond to a reduction of skin friction about 5.35% in flow without side wall effects. In case of a drag increase this relation holds as well. The estimate is limited to moderate changes of the skin-friction to exclude a modification of the global flow field; its validity is restricted to a channel with an aspect ratio of 12. If an increase of the aspect ratio is assumed to mainly extend the region of two-dimensional flow in the center, the concept can be adopted to predict a general function $\Delta\tau_w = f(AS)$. For $AS > 5$ the estimation can be formulated:

$$\begin{aligned}\Delta\tau_{w,2D} &= (1 + (\overbrace{0.5\frac{1}{AS}}^{\text{sidewalls}} + \overbrace{0.16\frac{1}{AS}}^{\substack{\text{corner}\\\text{region}}} + \overbrace{0.19\frac{1}{AS}}^{\substack{\text{spanwise}\\\text{variation}}}))\Delta\tau_{w,ch} \\ &= (1 + 0.85\frac{1}{AS})\Delta\tau_{w,ch}\end{aligned} \tag{5.9}$$

In Figure 5.21 equation (5.9) is plotted for $AS > 5$. The effect of the sidewall is quite large for $AS < 10$. An aspect ratio of 20-30 is desirable, since the effect then is almost negligible and the advance of larger aspect ratios is small. With increasing Re slight development towards a two-dimensional velocity distribution is indicated by the results for the distribution of spanwise centerline velocity and wall friction. Thus for much higher Re than in the present study the change in the distribution of τ_w might be too large to assume adaptability of the approach.

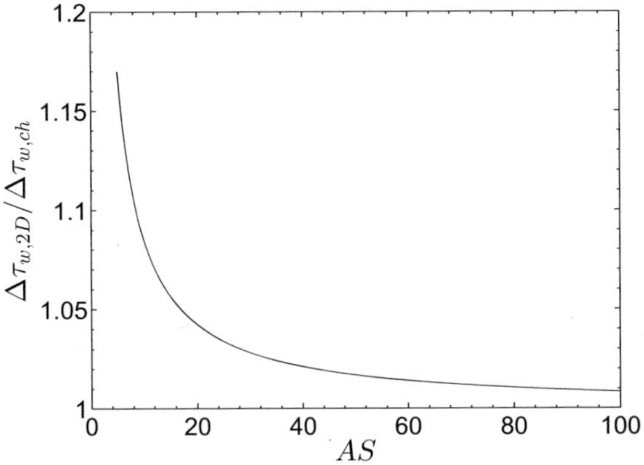

Figure 5.21: Estimated influence of a channel experiment on $\Delta\tau_{w,2D}/\Delta\tau_{w,ch}$ in dependence of AS.

The underestimation of the change in $\tau_w/\tau_{w,0}$ is not the only difference to an ideal two-dimensional flow field. On the main portion of the top and bottom walls, where a drag reduction application works most effectively, the displacement of fluid from side walls to the tunnel center leads to an higher effective Reynolds number than Re_b indicates. The latter is calculated as the integral value for the entire cross section for the experimental study. Therefore Re_b does not represent exactly the same quantity as for an ideal two-dimensional flow corresponding to the bulk value in numerical simulations. As a result, the higher effective bulk Reynolds number $Re_{b,eff}$ of the experimental study slightly shifts the results of $\tau_w/\tau_{w,0}$ to lower Re_b. The magnitude of the shift in Re_b can estimated from comparison of the numerical data with the Pitot tube measurement, which contrary to the Preston tube results is very accurate ($\delta U \ll 1\%$). The spanwise centerline velocity is assumed to indicate the local Re_b for a vertical plane of the channel, since the distribution of τ_w is very similar to that on the spanwise centerline velocity. From a comparison of the velocity at the centerline of the DNS $U_{c,DNS}$ with the Pitot tube measurement in the channel center U_c, an increase of $\approx 2\%$ is obtained in the experiment. If only the region

70% around center of the spanwise distribution are considered, the mean value is ≈ 1% higher. The resulting shift in Re_b will be very small.

Generally, for $AS = 12$ the shift in Re_b as well as the amount of underestimating $\Delta\tau_w$ is smaller than the measurement accuracy commonly achieved in previous experimental or numerical investigations.

5.4.4 Summary of the reference measurements

Since this Chapter included a number of different aspects, the essential findings are summarized here because of their fundamental importance:

- For naturally developing flow, the transition occurs above $Re_{trans} = 6000$; a fully developed turbulent state on the downstream section is not found until $Re > 12000$.
- With the artificially triggered transition a tripping level of 13% is found to be appropriate. A fully turbulent flow was indicated for $Re > 5000$ on both downstream sections and for $Re > 4000$ on the final section.
- A development length of $L_{E,turb} = 70 - 80H$ is needed for the turbulent flow. This means, on the first half of the channel length a fully turbulent, but not a fully developed turbulent flow state can be achieved.
- For both flow states a hydraulically smooth surface is confirmed for the entire range of Re_b.
- The data scatter is small for large range of Re_b and generally in good agreement to maximum scatter predicted in Chapter 4.
- The influence of the deviation of the wall shear stress distribution from an ideal, two-dimensional flow field $\Delta\tau_{w,}/\Delta\tau_{w,0}$ is approximately 7% for the channel of the present investigation with $AS = 12$. A correction function to compute the difference has been derived.

6 Riblets

In this chapter, the previously described wind tunnel facility is used to investigate the effect of riblets on the flow. This investigation has two goals: The first goal is to demonstrate the flow control capabilities of the new facility. The high quality data obtained with riblet-structured channel walls enables comparison of the passive flow control results with the existing literature.

The second goal is to study the streamwise flow development in a very long channel section with riblet surfaces, which was not possible in other investigations before with such high accuracy. Additionally, the limitations in comparability of the present results with the available literature are discussed and the implications on the drag reduction prediction are highlighted.

6.1 Introduction

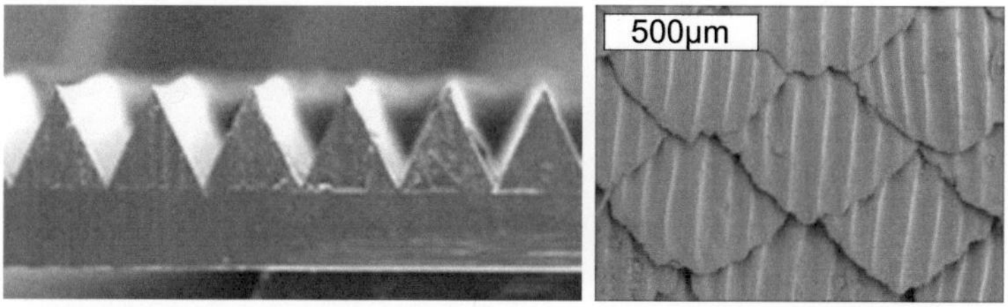

Figure 6.1: Riblets for technical applications produced by 3M (left) (University of Minnesota, www.safl.umn.edu) and structure of shark skin (right) (www.maritimejournal.com).

The term riblets refers to flow-aligned fins on the surface enabling a reduction of skin friction up to 10% (Bechert *et al.*, 1997). Figure 6.1 exemplary

shows highly magnified pictures of the small scale fins on a technical fabricated structure and shark skin, which is a well known naturally occuring example for riblets.
The success of riblets, in contrast to almost all other ideas for reducing skin friction drag, is the achieved breakthrough in technical applications. For instance riblets are used to enhance the performance of swimming suits. In the recent years the application on the hull of sail boats especially for the famous yacht races lead to a renaissance of the attention to the structures. The structures have been tested on water and wind turbines (Sareen *et al.*, 2011). Futhermore, a potential to reduce skin friction on high Reynolds number flows was demonstrated in test on jet airplanes and aircraft (McLean *et al.*, 1987; Viswanath, 2002; Walsh & Sellers, 1988). Despite their effectiveness, economic considerations showed that the required effort for the application and maintenance of the structures is not cost-effective. Thus, optimization is still necessary for a profitable application on civil aircraft.

Walsh (1990) describes two fundamental approaches that led approximately simultaneously to the design of the structure. On the one hand there have been several scientific approaches to influence near wall vortices in a turbulent boundary layer resulting in a surface with streamwise orientated fins (Walsh & Weinstein, 1978), (Kennedy *et al.*, 1973). On the other hand the observation of the morphology of shark skin showed small fins that are arranged in flow direction (Reif, 1978), (Reif & Dinkelacker, 1982). The function of these little fins is attributed to the same mechanism of skin friction reduction as the similar shaped technical surface.

6.2 Physical principle

The drag reduction mechanism of riblets is based on a hampering of the interaction of vortex structures in a turbulent flow close to the wall. Although the complexity of the process is not yet understood in all details, the development of the flow on a riblet surface can be characterized. A detailed summary of the actual state of science is provided by Dean & Bhushan (2010). The following explanation outlines their comprehensive description.

The high friction drag of turbulent flows originates from an increase of momentum transfer in the wall normal direction, especially in the near wall region. This transfer of momentum to the wall changes the shape of the mean velocity profile. In consequence the velocity profile becomes fuller with a higher gradient $\partial\overline{U_1}/\partial x_2$ at the wall. Following the concept of roughness in a turbulent boundary layer any increase of the wetted surface enhances the shear stress. Consequently the presence of riblet structures, which are increasing the surface, should result in an increase of friction drag. Without the usual statistical consideration of turbulence, a closer examination of the fluctuating flow reveals recurring structures in the near-wall region. These are vortices with the axis of rotation orientated in flow direction. The interaction of the quasi-streamwise vortices with the surface contribute to a large extent to the momentum transfer. The position of the vortices is varying due to their movement in spanwise direction. Owing to the transverse motion, adjacent vortices can collide, resulting in a fast ejection of fluid from the near wall region into the outer layer (Kline *et al.*, 1967). This bursting motion away from the wall, the transverse movement of quasi-streamwise vortices and the interaction of these events cause additional momentum transfer and thus play a major role for the amount of friction drag in turbulent wall bounded flows. One key to reduce the friction drag is to limit these processes. One possibility to reduce such effects is the use of specifically designed fins on the wall surface, i.e. riblet structure. Riblets will act effectively in reducing friction drag, if the hampering of vortex interaction outruns the drag increase due to the enlarged surface area.

The general mechanism is to prevent the quasi-streamwise vortices from directly attaching to the wall. If the vortices only interact with the protruding riblet tips, the near wall region between the riblets is shielded. In contrast to a smooth surface the vortices can not directly attach to the wall. The cross-fluctuation in between the fins of the riblets decrease, resulting in a successive lower shear stress. Although friction increases in vicinity of the tips the average skin friction is reduced.

The effectiveness of riblet drag reduction depends on the scaling of the structure, which needs to correlate with the dimensions of the quasi-streamwise vortices. Kline *et al.* (1967),Walsh (1982, 1990) (experimental) and Choi *et al.* (1993) (numerical) indicate vortex diameters in the range of 35 to 50 wall units. As the dimensions of the vortices depend on the Reynolds number, the riblet spacing s can only be optimized for one selected design point. For large s the vortices come close to the wall in between the fins and interact with the increased surface area with the outcome of higher drag than smooth surface. The visualization of vortex structures in a wind tunnel experiment by Lee & Lee (2001) (Fig. 6.2) shows a comparison of the behavior of streamwise vortices on a structured wall for different Reynolds numbers. The interaction of the streamwise vortices and the surface is depicted for a riblet spacing $s^+ = s/\delta_\nu = 25$ and for a scaling of $s^+ = 40$ for semi-circular riblets as used by Walsh (1983). The dimensionless spacing s^+ is obtained by normalizing s with the viscous length δ_ν, which is Re dependent (see Chap. 2.1.2). In the case of drag reduction the riblets limit the spanwise movement of the vortices and induce secondary vortices (Lee & Lee, 2001). In turn a smaller level of the velocity fluctuations was detected. For the drag increasing case at the higher Reynolds number, the spacing is larger than the diameter of the quasi-streamwise vortices. The vortices are observed closer to the wall and mostly stay in between the riblet tips. Lee & Lee (2001) explain the increase of friction drag with the down wash of high speed flow from the outer layer to the larger wetted surface of the riblet structure (Lee & Lee, 2001).

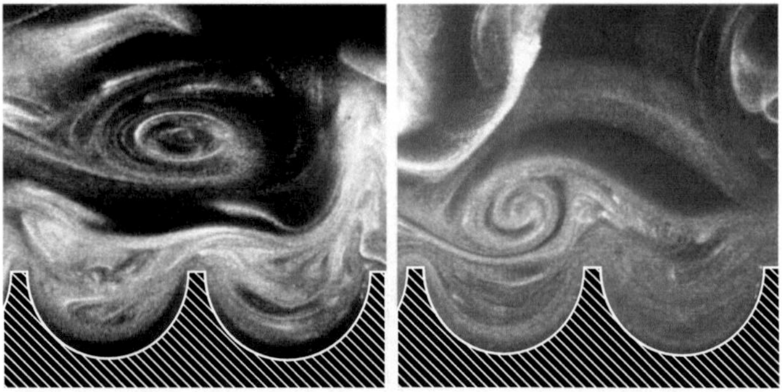

Figure 6.2: Flow visualization of streamwise vortices on a plate with circular riblets at 3m/s (drag decreasing) and 5m/s (drag increasing) from Lee & Lee (2001). Reproduced with kind permission from Springer Science and Business Media.

To quantify the relation between the riblets geometry and the amount of skin friction reduction Bechert & Bartenwerfer (1989) introduced the concept of protrusion height difference. The protrusion height is defined as the difference of vertical shift of the flow origin to the tip height of the riblets (Fig. 6.3). This parameter can be calculated for the flow parallel h_{ps} and in cross direction h_{pc} to the fins. The difference of the two flow

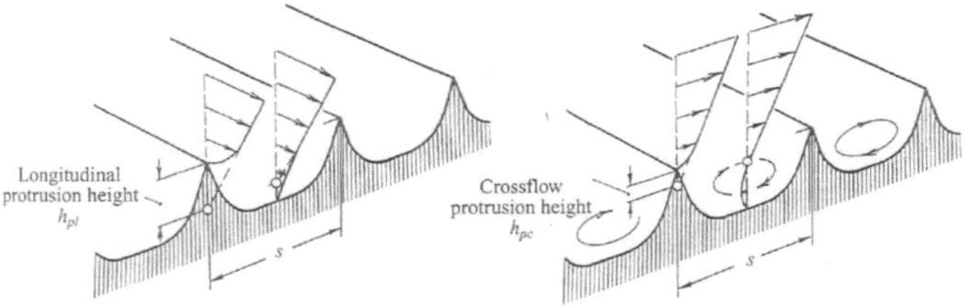

Figure 6.3: Concept of protrusion height difference taken from Bechert *et al.* (1997). Reproduced with permission.

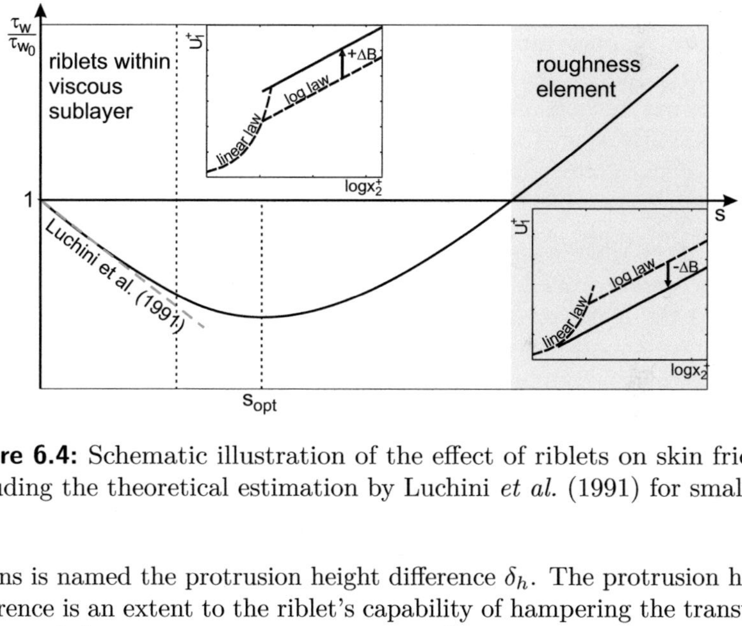

Figure 6.4: Schematic illustration of the effect of riblets on skin friction including the theoretical estimation by Luchini *et al.* (1991) for small s.

origins is named the protrusion height difference δ_h. The protrusion height difference is an extent to the riblet's capability of hampering the transverse motion of the quasi-streamwise vortices. Since this concept is based on viscous theory, it is only valid if the riblets are embedded in the viscous sublayer. With increasing s^+ the riblets grow out of the sublayer and the concept of protrusion height difference loses its validity. Theoretical considerations by Luchini *et al.* (1991) support the concept of protrusion height. In oil tunnel experiments Grüneberger & Hage (2011) examined the effect of small riblet spacings and found the effect of the riblets linearly decreasing with the riblet spacing for $s < s_{opt}$, which is in accordance to Bechert's and Lucchini's theories (compare Fig. 6.4).

There is a number of experimental and numerical studies considering the drag reduction mechanism of riblets from differing points of view. Exemplarily Lee & Choi (2008) pay attention to the presence of spanwise vortices in their PIV experiments. Spanwise vortices are in close relation to the mechanism of sweeps and ejections. Their finding is a modification of the spanwise vortices in the near wall region. Especially the number of vortices with a rotation in counter direction to the near wall flow, which are responsible for the drag increasing ejection event, is significantly reduced. In numerical investigation Choi *et al.* (1993); Garcia-Mayoral & Jiménez (2011) were able to analyze the flow field surrounding the riblets and to identify vortex structures in the vicinity of the riblet tips.

The mentioned observations from varying points of view consider different aspects and show different mechanisms of the riblets effecting the flow. These are not in contradiction, but contribute to an overall picture, viewed from differing angles. Today's discussion on the mechanisms is the question, which are the underlying mechanisms and which are an implication of those.

Regarding the riblet's impact on a turbulent boundary, three main parameters have been identified: The spanwise distance between the fin tips, which is named the riblet spacing s, the total riblet height h from the lowest point in valley between the fins to the tips and the riblet shape. s and h can be transformed to nondimensional quantities as introduced in Chapter 2.1.2:

$$s^+ = \frac{s}{\delta_v}, \; h^+ = \frac{h}{\delta_v} \tag{6.1}$$

For the optimum geometric parameters Bechert *et al.* (1997) give a spacing of $s^+ = 16$ to $s^+ = 17$ and ratio $h/s = 0.5$, which is in general agreement to other literature (Dean & Bhushan, 2010; Walsh, 1990). One should denote that this specification varies for different geometries since often both critera cannot be kept simultaneously. In the case of sawtooth shape the ratio is a function of the tip angle α. For the extensively studied geometry with $\alpha = 60°$, which has been produced industrially by 3M, the ratio is $h/s = 0.87$. It can be summarized that for optimum effectiveness the riblet tips slightly reach into the buffer layer $h^+ > 5$, but are still far from the logarithmic layer. The optimum rib spacing is approximately half the diameter of a quasi-streamwise vortex. Schematically the dependence of $\tau_w/\tau_{w,0}$ on the riblet spacing is summarized in Figure 6.4.

The optimal design for riblets is an arrangement of infinitely thin vertical blades. This geometry was realized in good approximation by Bechert *et al.* (1997) and led to the highest skin friction reduction of 9.9% verified in experiments so far. Luchini's analytical considerations confirmed the blade shape as the most effective (Luchini, 1993). Although the blade riblets showed highest effectiveness, their practical use is restricted due to their vulnerability to mechanical loads. Walsh (1990) suggested to further study triangular riblets with a flat interspace. (Bechert *et al.*, 1997) reported a decrease of 8% in skin for such a configuration, which is quite close to the blade riblets. This shape appears to be a good compromise between durability and performance. Further investigations with this geometry were conducted by Grüneberger & Hage (2011).

During the last 20 years the attempt was made to improve riblets effectiveness by modeling the shape in streamwise direction. Length, arrangement and flexibility of the structures were varied to simulate natural shark skin (Bechert *et al.*, 1997). Additional effort was put on the modification of riblets to a wave like structure to induce a supplemental spanwise mechanism (Grüneberger & Hage, 2011). In experimental investigations these modifications did not show benefits compared with the standard two-dimensional geometry of riblets. In contrast the drag reduction was even less.

The effectiveness of riblets could be verified up to the region of transonic flows (Viswanath, 2002; Walsh, 1990), where a slight increase of the performance was indicated. In addition, riblets have also been tested in laminar and transitional flow. In the first case some few investigations report small drag reduction (Fulachier *et al.*, 1987; Liandrat, 1988). In studies at the Langley Research Center no changes of the skin friction could be found (Walsh, 1990), which is in accordance to theoretical study of Choi *et al.* (1991). Thus Walsh (1990) condiders the drag reduction reported in aforementioned studies to be related to uncertainty of the detection procedere. Today it is common sense that riblets only reduce drag in turbulent flows. In laminar flow the complete mechanism described for turbulence is not effective due to the absence of quasi-streamwise vortices. The structures will act as roughness elements and increase the wetted surface.

In transitional flow, sources in literature are not in complete agreement, but a transition delay has not been definitely proofed. For instance a study on an airfoil on glider plane conducted by Quast is reported in

Bechert (1988), where initially promising reduction of skin friction change is stated to be related to the control of a separation bubble. In the theoretical work of Lucchini (1995) a modification of the instabilities in transition could be found, but a clear statement, whether this would lead to a significant reduction of friction drag was not made. Hence, the present study concentrates on 2D-shaped riblets in turbulent flow conditions.

6.3 Experimental study of riblets

6.3.1 Experimental investigation - State-of-the-art

Within the large number of articles on this topic two outstanding sources of comprehensive experimental work are found: The investigations of Walsh and the studies of Bechert. In the early 1980s Walsh's pioneering work yielded a comprehensive characterization of the effect of riblets (Walsh, 1982, 1983; Walsh & Lindemann, 1984). Within a short period many different geometries have been examined. Bechert's investigations are outstanding especially because of the high accuracy measurements (Bechert *et al.*, 1992). Both authors extensively studied the effect of the riblet geometry on drag reduction. A summary of those results is given in Bechert *et al.* (1997) and Walsh (1982) respectively Walsh & Lindemann (1984) .

Walsh (1983) calculated the height of the riblets for cruise flight ($Ma = 0.8$) at an altitude of 12.2 km and a Reynolds number $Re_x = 10^6$ as $h = 0.076\text{mm}$. Difficulties of riblet fabrication and the influence of apparent imperfections originating from manufacturing processes result in significant experimental uncertainty, as shown by Walsh & Lindemann (1984). Thus Walsh points out the challenge which still concerns a lot of engineers and scientist: The application and high quality fabrication of the small scale structures on large surface area.

In laboratory experiments the difficulty of fabrication can be reduced, using liquids as flow medium. Bechert *et al.* (1992) utilize the favorable ratio of dynamic viscosity to density of oil compared to air ($\upsilon_{oil} = 1.20 \times 10^{-5} m^2/s$). The riblet size at equivalent Reynolds number grows from few tenths of a millimeter in air to few millimeters in oil.

For the purpose of riblet testing a new oil tunnel facility was designed and built in Berlin (Bechert *et al.*, 1992), possessing very high measurement

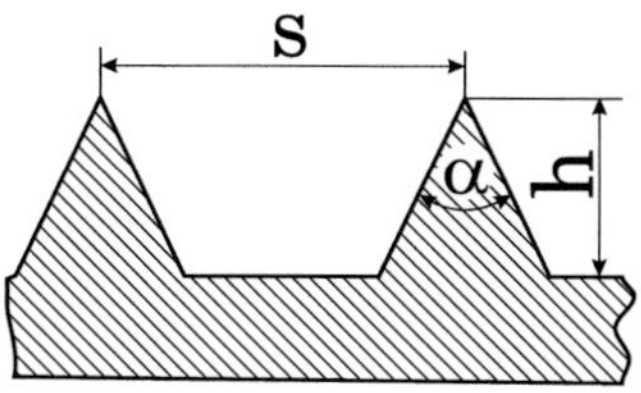

Figure 6.5: Schematic sketch of riblets type with trapezodial geometry.

accuracy. While Walsh specifies a value of ±1% accurary for the determination of friction change in his wind tunnel facility, which partly seems optimistic regarding the presented results, Bechert attains ±0.3%.

It has to be denoted that oil and water tunnels have a disadvantage. If the streamwise length of the riblet surface is considered as a multiple of the riblet height, wind tunnels can be equipped with a much longer riblet surfaces, which is necessary to study the flow development on the structures.

6.3.2 Design and fabrication of the structures

Since at the time of the present study riblet surfaces have not been available on the market, it became inevitable to manufacture the structures. To obtain a preferably high amount of difference in friction between the smooth reference and the riblet surface, a triangular geometry with flat interspaces was chosen (Fig. 6.5). This design promises the optimum compromise between manufacturing capabilities and a high reduction of skin friction (Grüneberger & Hage, 2011). The calculation of the geometric parameters is carried out following Bechert *et al.* (1997), who extensively studied this riblet shape. With decreasing tip angles $(60°, 45°, 30°)$ an increase in effectiveness was found. Accordingly a geometry with a ratio $h/s = 0.5$, $s^+_{opt} = 16.5$ and the sharpest contrivable tip angle was selected $(53.5°)$ for the present study. The design Reynolds number was chosen to $Re_b = 11700$ in order to guaranty fully turbulent flow, even in a wide range below the optimum $(s^+ < s^+_{opt})$. Due to the agreement of the results in Chapter 5.2 with Dean's formula, the calculation of τ_w is based on this empirical correlation for the riblet design. Table 6.1 presents the caculated geometry for the selected design point.

The structures were manufactured in a large milling machine at the Fraunhofer Institute for Production Technology of the RWTH Aachen. Special diamond cutting tools were custom made by Mössner in Pforzheim. The dimensions of the tools slightly deviate from the design values which led to minimal differences of the geometry (Tab. 6.1). Two plates with a structured surface area of 1500mm x 300mm were produced with micrometer precision. A cross section view of the riblets is presented in Figure 6.6.

Planarity is guaranteed by the use of 15mm aluminium baseplates. A thick layer of plastic material is screwed on the baseplates and the surface has been milled to planarity before in the final step the riblet structure was fabricated. Due to machinability and minor abrasion of the tool it has been decided to use plastic material for the riblets. Indeed a single cutting tool did the job and thus ensured the structure to be identical over the entire surface. Especially the crucial tips of the riblets could be realized in an excellent finish.

From the viewpoint of a channel experiment also the intersection between the smooth plates and riblet plates needed to be considered. According to Bechert *et al.* (1997) an abrupt begin of the riblet surface should not alter the flow at the onset of the riblet plate. However, due to the small dimensions of the present channel a change of the cross sectional area due to the presence riblets is not negligible. Therefore the cross section area is kept constant by increasing the channel height to compensate for the blockage arising from the protruding riblets. This was realized by a protruding frame on both sides of the riblet plate. In Figure 6.7 the step of 70μm at the contact surface of the side wall is shown.

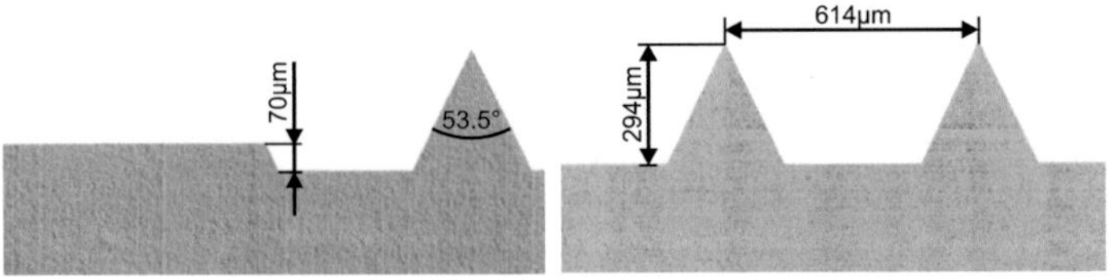

Figure 6.6: Cross sectional view on trapezodial riblets manufactured at RWTH Aachen.

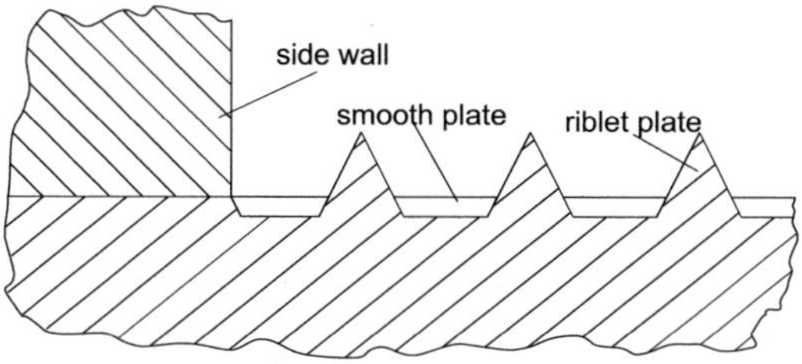

Figure 6.7: Detail of the intersection between the riblet, the smooth plate and the side wall. The direction of view is upstream.

Table 6.1: Designed and resulting dimensions the produced riblets

	s [μm]	h [μm]	tip angle [°]	Re_{opt}
Design	600	300	53.5	11700
Result	614	294	53.5	11400

6.3.3 Focus of the experimental investigations

Besides the necessity to demonstrate the effectiveness of the passive drag reduction mechanism in the developed facility with an high accuracy measurement system, there remain some contraverse discussed aspects concerning the flow development on riblet surfaces, on which the present experimentally setup might be able to shed some light. Therefore, the following paragraph depicts differences between the present set-up and the measurement technique from Walsh (1982) and Bechert *et al.* (1992). Considering a large number of literature data from different authors Walsh (1990) found those measured with a shear stress balances to be most accurate and reliable. Hence, Bechert as well as Walsh use shear stress balances in their studies. One disadvantage is that shear stress balances only provide integral drag values of the test plate. The streamwise development of τ_w is not resolved. Walsh (1982) investigates the effect of flow development by covering partly his test plate with tape to generate a smooth surface. The result shows a proportional relation between the riblet area and $\tau_w/\tau_{w,0}$. However, the accuracy of this procedure is questionable and a fine resolution cannot be achieved. The measurement of the pressure difference dp/dx in the present study automatically resolves the development in the flow direction depending on the number of pressure taps along the test plate.

Of particular interest is the question whether there is a streamwise distance required for reduction of skin friction and how long it is. Walsh collected some data concerning the flow development over a riblets surface and found inconsistent results (Walsh, 1990). While his own measurement indicated higher local skin-friction reduction, Gallagher & Thomas (1984) found the skin friction reduction to be less at the leading region of the riblets. Savill (1986) finds an immediate response to the presence of the structures, indicating a negligible development length. One difficulty is the adjustment of the begin of the riblets. On the one hand in Squire & Savill (1986, 1987) the leading edge affects the results. On the other hand Bechert *et al.* (1997) found no leading edge effect even for an abrupt begin of the riblet.

6.3.4 Arrangement of experimental setup

For the riblet investigation the channel is equipped with the 13%-tripping at $x = 0$ to trigger a turbulent flow. The structures are located in the downstream part of the test section $x > 98H$, where a fully developed turbulent flow is observed in the reference measurement (cf. Chap. 5.2). Experiments were conducted for a range of $Re_b = 3700 - 23500$. For this wide range two different flow rate instrument have been used: The orifice meter with the 60mm disk and the inlet nozzle with $D = 74$mm. During the measurement the pressure distribution over the total length of the test section has been read, but the data presented omit the first section, which is mainly affected by flow development on the 1000mm from the begin of the channel section. Results are presented starting at the sixth pressure tap located at 40H. The last pressure tap with a distance of 50mm to the end of the channel is also not shown, since the tap was obviously affected by the outlet for high Reynolds numbers (cf. Chap. 5.2). Thus the observed length is 1400mm on smooth surface, followed by the intersection of 100mm (50mm smooth, 50mm riblets) and a riblet length of 1200mm. In the following these sections are denoted "upstream" and "downstream". Both Baratron differential pressure sensors were used in combination with the Scanivalve multiplexer, where the first and the central tap have been read simultaneously by both sensors to check alignment of the sensors. The results are averaged over 180s for each pressure tap, resulting in 36min in total for one test series including 20 (22) readings.

6.4 Discussion of the results

To obtain an idea of the quality of the data and to present the effect of the riblets qualitatively, Figure 6.8 presents the friction coefficient in dependence of Re_b for the fully turbulent developed flow on downstream part of the test section for both surface conditions. Obviously the riblets are effective, since the c_f is lower for the structured surface at small and moderate Re_b. With increasing Re_b the distance between the two curves diminishes until finally the riblets act as additional roughness increasing the c_f. This observation is in agreement with the expected result.

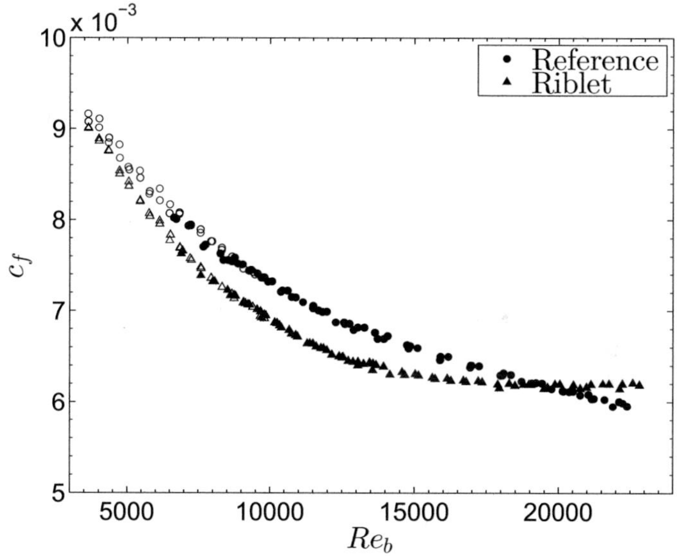

Figure 6.8: Comparison of friction factor c_f for smooth and structured surface in the downstream part of the test section. Open symbols indicate measurement with the orifice meter, filled symbols with inlet nozzle.

For further analysis we will first take a look at the distribution of the pressure loss (Fig. 6.10 and 6.11) in streamwise direction, shown pairwise for correlating Re_b. In the upstream section the surface is always smooth, while in the downstream part the surface consition is either smooth (right) and structured surface (left). Exemplarily a detailed look is provided in Figure 6.9 for $Re_b = 11300$. Scatter in the data is very low for both surface

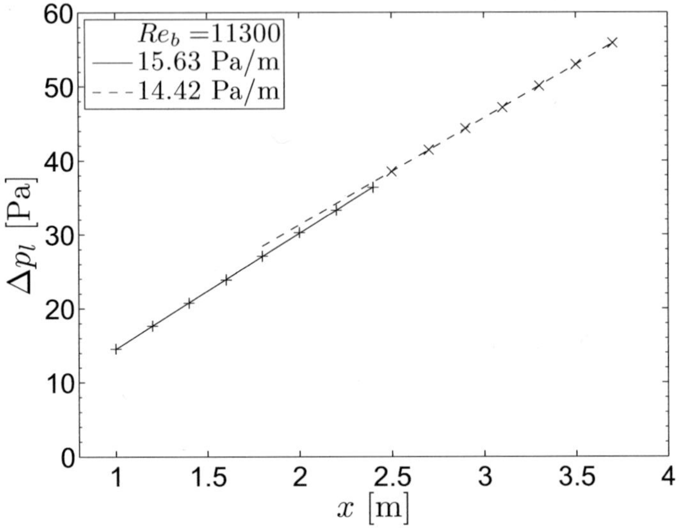

Figure 6.9: Pressure drop over smooth (+) and structured (x) surface along the channel.

conditions. The two lines show least-square-fits, one drawn through the eight taps of the smooth plate ($\Delta L = 1400$mm) and the other (dashed line) through the seven taps of the riblet plate ($\Delta L = 1200$mm). From figures Figures 6.10 and 6.11 a linear dependance of the static pressure gradient on the downstream position can be obtained. The residua R for the linear fit over seven considered pressure taps are $R > 0.99996$ for any of 120 data sets with riblet surface (and for the smooth surface as well). Neither an increase nor a decrease in riblets performance with respect to x can be found. Any development induced at the leading edge of the riblet plate is definitely vanished at 50mm which is equivalent to $2H$ or approximately $67h^+$. This observation is similar for all Re_b. For the lower Re_b with smaller pressure drop on the riblet surface than on the smooth plate as well as for Re_b=22230, where the dp/dx is larger on the riblet surface Walsh's proportional relation is confirmed. Surprisingly, the intersection between the smooth and the riblet surface, which is not taken into account for the calculation of $\tau_w/\tau_{w,0}$, exhibits a remarkable detail. The pressure taps next to the intersection, which were set in a closer distance of $100mm$ to each other, consisting of 50mm smooth and 50mm structured surface. Despite

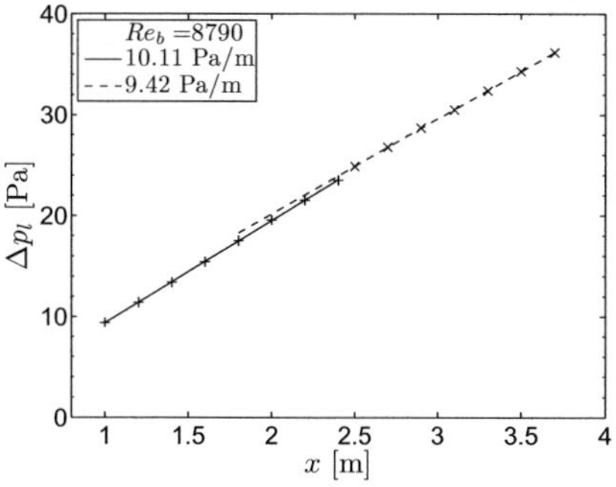

(a) $Re_b = 8790$

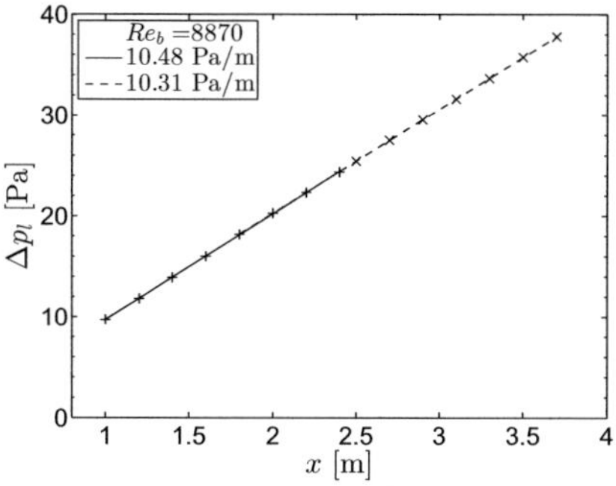

(b) $Re_b = 8870$

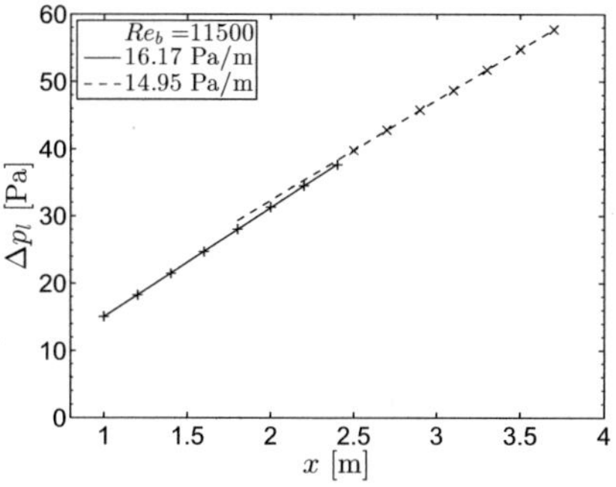

(c) $Re_b = 11500$

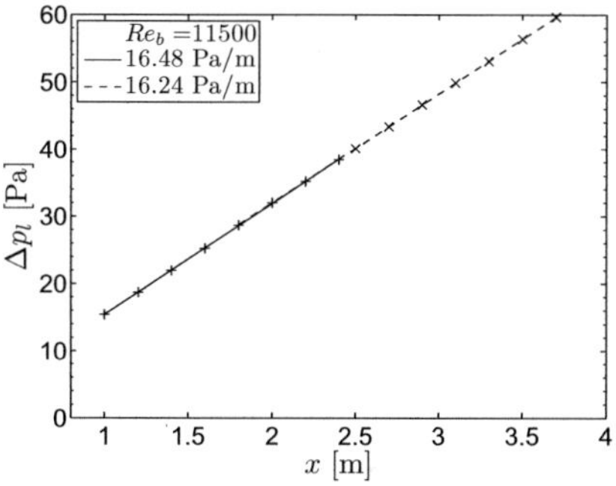

(d) $Re_b = 11500$

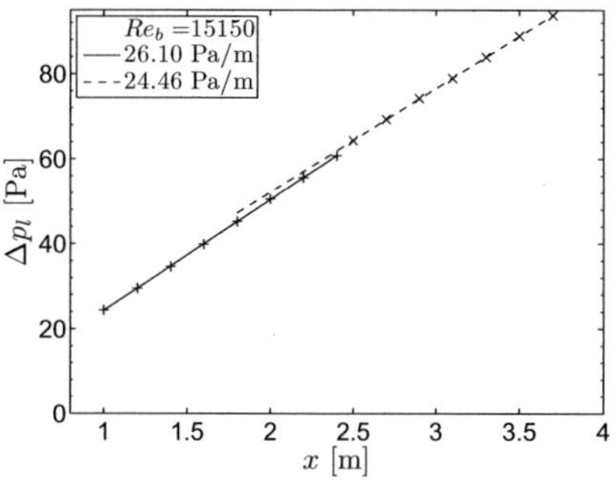

(e) $Re_b = 15150$

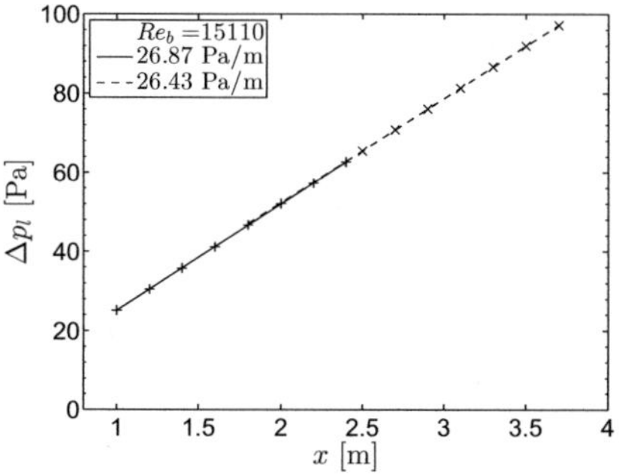

(f) $Re_b = 15110$

Figure 6.10: Comparison of pressure drop for completely smooth test section (right) and with the downstream section equipped with riblets (left) for different Re_b.

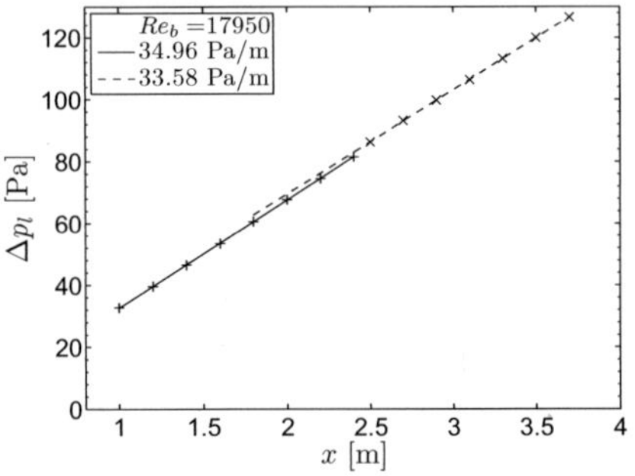

(a) $Re_b = 17950$

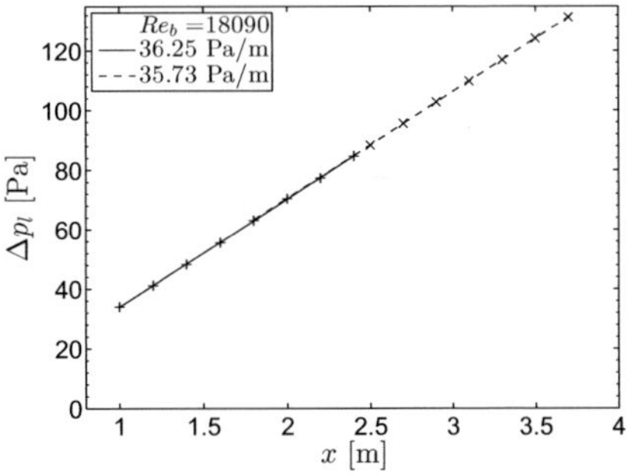

(b) $Re_b = 18090$

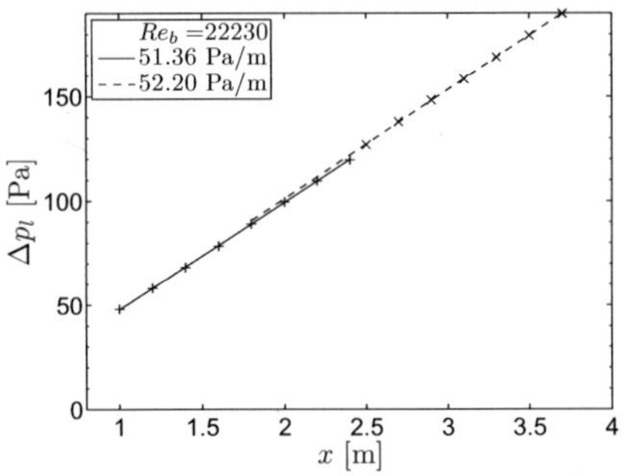

(c) $Re_b = 22230$

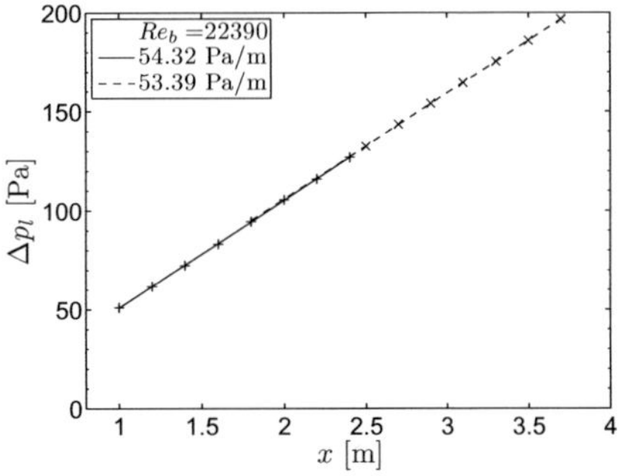

(d) $Re_b = 22390$

Figure 6.11: Comparison of pressure drop for the completely smooth test section (right) and the downstream section equipped with riblets (left) for different Re_b.

of the pressure loss of the average value of smooth and riblet surface, the data show a higher dp/dx than for the smooth plate. Since the reference data for the completely smooth channel (right column of Fig. 6.10 and 6.11) do not indicate any variation in dp/dx at the intersection, the effect must be related to the begin of the riblet plate. Obviously this observation is in contradiction to Bechert *et al.* (1997), who stated that an abrupt intersection between smooth the surface and riblets would not affect the results. Walsh (1982) reports a study by Sawyer & Winter (1987), where two configurations of the experimental set-up with a drag balance have been tested. In the first configuration the riblet structures were only located on the plate of the balance. In the second configuration the walls of the wind tunnel surrounding the drag balance were also with equipped with riblets. They reported no effect on the measurement results. Both authors could not resolve the development at the intersection, but refer to an integral drag value of the riblet surface, which may be affected slightly from the leading edge. This is considered to be the cause for the disagreement to the present examination. It needs to be mentioned here that nevertheless Bechert used optimized intersections for his measurement to exclude possible influence of the intersection. Details on this procedure are shown in appendix of Bechert *et al.* (1997).

In the present facility the pressure drop reacts quite sensitive even to small variations of the channel height H. At the intersection the effect of a backward facing step to keep the cross section constant cannot be distinguished from the influence of the protruding riblets. Note that the step is only 70μm on bottom and top wall, which in sum is far less than 1% of H. This indicates, that there is substantial contribution of the 225μm protruding riblets. In wind tunnels of conventional dimensions such a small step influences only insignificantly the mean velocity and thus Re_b. Therefore, this is not critical for the result of the shear stress τ_w. The differing reports in the literature might be influenced more by the details of the setup than by the riblets themselves. The result of the present study shows how critical small details in the design affect the results, if the quantities to be measured are small.

6.4.1 Strategy to determine the skin friction change

The decrease in skin friction can be determined by a comparison of $c_{f,0}$ for a smooth and c_f for a riblet surface in the rear part of the channel as it was shown in Figure 6.8.

$$\tau_w/\tau_{w,0} = c_f/c_{f,0} \tag{6.2}$$

The measurement uncertainty is then depending on two stochastic errors, namely the reproducibility of c_f, which can be assumed to be the same in the structured data. A Gaussian error propagation would lead to $\sqrt{2} \cdot \delta c_f$, resulting in an uncertainty of 0.4 to 1.3% depending on Re. The uncertainty is the result of the fact that the measurement needs to be conducted two times and the individual uncertainties of each measurement add (compare Chap. 4.3.2). The balance measurement by Bechert *et al.* (1997) directly compares the smooth and the structured surface and avoids this source of error. An additional disadvantage of the comparison of the downstream section data for different surface conditions is the presentation of the results. A comparison can only be made between two curve fits of the slightly scattered data for the smooth and the structured surface (Fig. 6.8). A direct comparison of single data points can hardly be drawn. Thus we will go a second way to calculate $\tau_w/\tau_{w,0}$ avoiding the reproduction error.

A direct comparison of the smooth and the structured surface is enabled, if both surface conditions are realized in one measurement. This can be achieved by comparison of the downstream riblet section with the upstream smooth section. The skin friction change can be calculated:

$$\frac{c_f}{c_{f,0}} = \frac{(\Delta p_l/\Delta L)}{(\Delta p_{l,0}/\Delta L_0)}. \tag{6.3}$$

However, from the reference measurement we know that the flow is fully developed for a channel length of $x \geq 80$H. So the flow development slightly influences the smooth upstream section of the channel. In consequence the smooth upstream section exhibits little more pressure loss and the comparison would to lead an apparent skin friction reduction, originating from incomplete flow development on the upstream section. Thus a direct comparison can only be drawn, if the degree of this "under-development" can be determined. Therefore in Figure 6.12 the ratio of the skin friction on

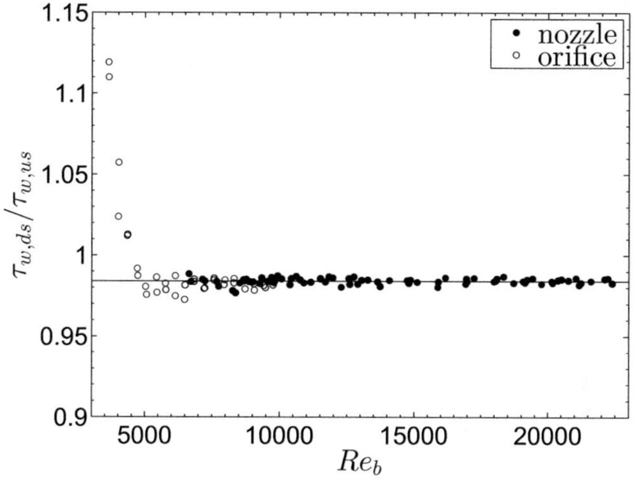

Figure 6.12: Ratio between the skin friction on the downstream and the upstream plate for a smooth reference channel for $Re_b = 3700 - 23000$.

the upstream and the downstream section for the complete smooth channel is shown. A constant relation between the pressure loss on the two sections is obtained for $Re_b > 5000$ (Fig. 6.12), where the flow is fully turbulent. If it is possible to show, that the results from the upstream part are exactly the same independent of whether a smooth or a structured plate is located downstream, a constant factor k_d can be determined (Eq. 6.4), which is used to correct the result of $c_f/c_{f,0}$:

$$\frac{(\Delta p_{l,0}/\Delta L_0)_{\text{downstream}}}{(\Delta p_{l,0}/\Delta L_0)_{\text{upstream}}} = k_d. \tag{6.4}$$

Then dp/dx for downstream section with riblets and the smooth upstream section can be directly compared:

$$\frac{c_f}{c_{f,0}} = \frac{\frac{(\Delta p_l/\Delta L)}{(\Delta p_{l,0}/\Delta L_0)}}{k_d}. \tag{6.5}$$

The constant is determined from Figure 6.12 to $k_d = 1.016$.

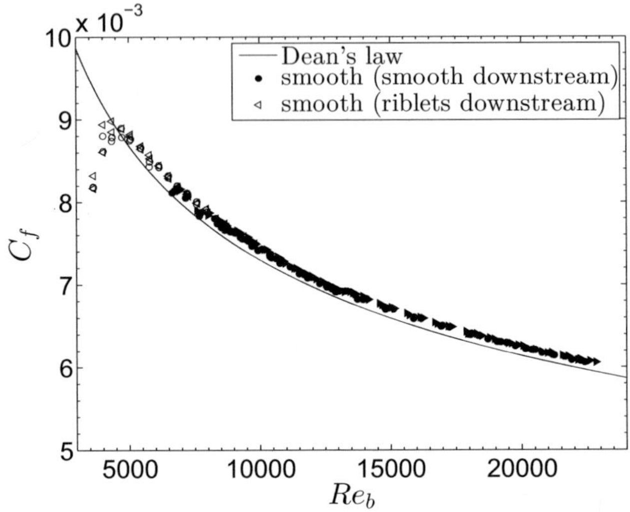

Figure 6.13: c_f over Re_b on the upstream section with the smooth (circles) or riblet surface (triangles) on the downstream section. Open symbols indicate measurement with the orifice meter, filled symbols with the inlet nozzle.

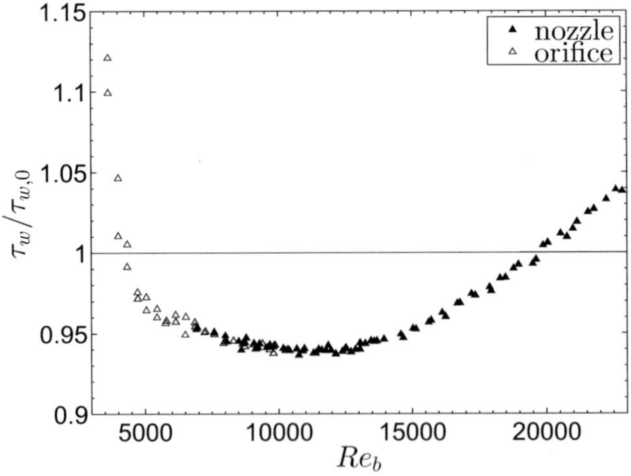

Figure 6.14: Distribution of $\tau_w/\tau_{w,0}$ in dependence of Re_b.

Considering the uncertainty in a distribution of $\tau_w/\tau_{w,0}$ over Re_b the reproducibility of the flow rate measurement will be the dominating factor in relation to the small errors originating from the pressure drop measurement. The reproducibility of the flow rate will lead to a horizontal scatter of the data points, whereas the uncertainty originating from the pressure drop reading is very small, since it depends only on the reproducibility of the high accuracy sensors from MKS (see Chap. 4.3.2).

The validity of this procedure is verified by the comparison of the friction coefficient on the smooth downstream plate for the case of the reference and the riblet measurements in Figure 6.13. The curve fits of each data set demonstrate a minimum difference between both test series over the complete range of Re_b. For smaller Re_b a slight difference of 0.2% is obtained, which is hardly visible in the diagram and lies clearly within the measurement accuracy. Note that the difference between the flow rate instruments in a range of $6700 < Re_b < 7500$ is due to the lower limit of validity of the calibration of the nozzle (filled symbols) and is also repeatable for both conditions of the test section.

For the use of this strategy to determine $\tau_w/\tau_{w,0}$ it is fundamental to have a fully turbulent flow at the beginning of the upstream plate, which is not the case for small Re_b as can be seen in Figure 6.12. The benefit of this strategy with respect to the confidence of $\tau_w/\tau_{w,0}$ is limited to $Re > 7000$, where the scatter in the data is small and there is sufficient clearance to the transitional regime. The result of $\tau_w/\tau_{w,0}$ is presented in Figure 6.14. Indeed for $Re_b > 7000$ a small scatter in the range of $\pm 0.5\%$ to $\pm 0.2\%$ is observed, while for smaller Re_b the scatter is higher. At $Re_b < 4800$ the flow along the upstream plate is still transitional and the data indicate a lower drag on the downstream plate due to partly laminar flow. This was also seen in Figure 6.12. Therefore, the following discussion will not take into consideration measurements for $Re_b < 5000$.

6.4.2 Change of the skin friction drag

A direct comparison of data from the study of trapezoidal riblets from Bechert *et al.* (1997) with the present results is given in Figure 6.15. In general, agreement with the data of Bechert *et al.* (1997) is found. The graph for the used 53.5°-riblets lies in between the curves of 45° and 60°-Riblets from Bechert's data. The optimum spacing $s^+_{opt} = 16.5$ is exactly the same and the maximum drag reduction of 6.0% represents the expected value for 53.5°-riblets. The trend of decreasing reduction of τ_w for $s^+ > s^+_{opt}$ is reproduced and the region of inversion from drag decrease to increase at $s^+ = 25 - 30$ is reproduced. With regard to the quality of the data, the maximum scatter is similar to the oil tunnel data for $s^+ > 11$. On the first view only one detail seems to differ from Bechert's data: For $s^+ < s^+_{opt}$ the gradient is steeper in the present study. Initially this was thought to be related to the vicinity to the transitional state of the upstream part of the smooth section, but an analysis comparing the downstream sections for the riblet measurement and the reference case led to the same result. However, there may be another reason. More insight is obtained by a detailed look on the measurement procedure in Bechert *et al.* (1997).

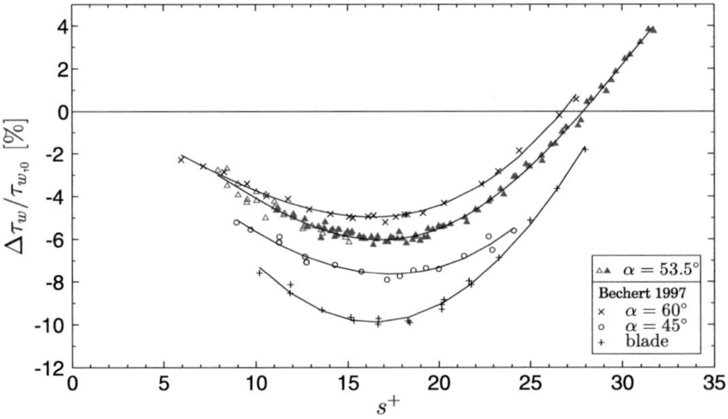

Figure 6.15: $\Delta\tau_w/\tau_{w,0}$ in dependence of s^+ in comparison to data adopted from Bechert *et al.* (1997).

The sketch on left side of Figure 6.17 depicts the cross section of the tunnel used by Bechert *et al.* (1997) with a shape of a rectangular duct. The test plates are located at the side walls and cover only the center part, which is used to detect τ_w via a shear stress balance. The distribution of τ_w is shown for two configurations: For an open tunnel configuration with a free surface and a closed channel with adjustable pressure gradient using a plate inclined in the streamwise direction. In the open top configuration the tunnel has a pressure drop in streamwise direction. The configuration with an adjustable top wall enables a zero pressure gradient over the length of test plates and was used for riblet testing in Bechert *et al.* (1997). Since the measurement in the present work is based on pressure drop, the open configuration is better comparable to the channel flow. Bechert *et al.* (1997)

present a comparison of measurement data with both configurations, which is shown in Figure 6.16. If the data from the open top configuration are considered, the gradient for $s^+ < s^+_{opt}$ appears to be similar to the data of the present study. It is assumed that the difference in the measurement for small s^+ is caused by the pressure gradient. However, this still cannot explain the steeper gradient. From the theoretical point of view a linear behavior matching the coordinate origin according to Luchini's theory is expected. The open tunnel configuration in the study of Bechert provides significantly more uniform spanwise distribution of τ_w, which should better agree to the theory (for further details see the following section). But the opposite is the case. How the different spanwise distribution of τ_w influences the drag reduction for riblets with $s^+ < s^+_{opt}$ remains open.

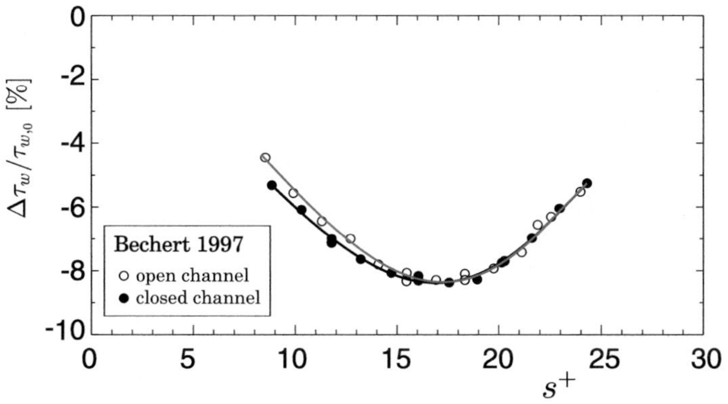

Figure 6.16: Distribution of τ_w over Re in dependence on tunnel configuration redrawn from Bechert *et al.* (1997).

6.4.3 Influence of spanwise velocity distribution

As discussed in Section 5.4 the investigated channel flow exhibits a nearly uniform velocity distribution in the center part of its cross section. Close to the side walls the shear stress deviates from a two-dimensional distribution. This was examined in Chapter 5.4. The presence of the unstructured side walls causes additional friction. In consequence one could expect the drag reduction with the same riblet geometry to be less than in experiments in other facilities. Nevertheless, even without the correction for the flat duct (Eq. 5.9) the agreement is quite good.

In the following some general aspects affecting the measurement are discussed based on a comparison of the present set-up with the experimental facilities used by Walsh (1982) and Bechert *et al.* (1997).

In Walsh (1982, 1983) and Walsh & Lindemann (1984) a wind tunnel with a rectangular cross section of 279mm x 178mm and a length of 914mm has been used. This cross section geometry with an aspect ratio of 1:1.5 does not provide a smooth lateral shear stress distribution. With dimensions of 279mm x 178mm the test plate uses the complete lateral spacing of the tunnel including the secondary flow dominated region close to the corners which significantly influence the velocity distribution.

Bechert *et al.* (1997) give an comprehensive description of their facility in Bechert *et al.* (1992). In the appendix to their article, Bechert *et al.* (1997) also present the test section in riblet testing configuration in detail. With a cross section of 250mm x 850mm an aspect ratio of approximately 1:3.5 is achieved. The drag balance is located in the middle part of the 850mm long side wall with a width of 400mm. Advantageously in this arrangement the sidewall effects can be eliminated to some degree. Otherwise the aspect ratio of 1:3.5 is not large enough to guarantee a two-dimensional velocity distribution in the middle part. The lateral distribution of τ_w on a flat and a riblet plate was determined with a Preston tube in the oil tunnel. Figure 6.17 shows the results for open channel and zero pressure gradient configuration. In both cases the skin friction is higher in the middle and decreases in lateral direction. For the zero pressure configuration, which has been used for the riblet tests, the profile becomes asymmetric and the wall shear stress variation of 20% is much higher. The variation in flow direction is not shown, but it can be assumed that the deceleration of the fluid needed to achieve $dp/dx = 0$ will also cause a variation of shear stress in flow direction. Note that the difference in drag reduction for low Reynolds number presented in Figure 6.16 is considered to be a pure result from differing tunnel configuration, meaning the τ_w distribution, in Bechert *et al.* (1997).

In summary, measurement results in all three facilities are affected by a non-constant shear stress distribution which can slightly influence the maximum amount of drag reduction. As indicated in Figure 6.16 the deviation might be larger for $s^+ < s^+_{opt}$ depending on the chosen experimental procedure. For a comparison of experimental with numerical 2D data these should be kept in mind.

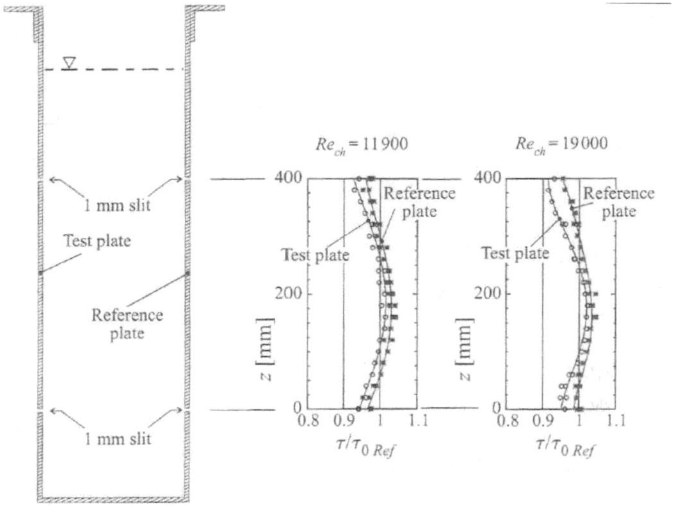

(a) Open tunnel configuration

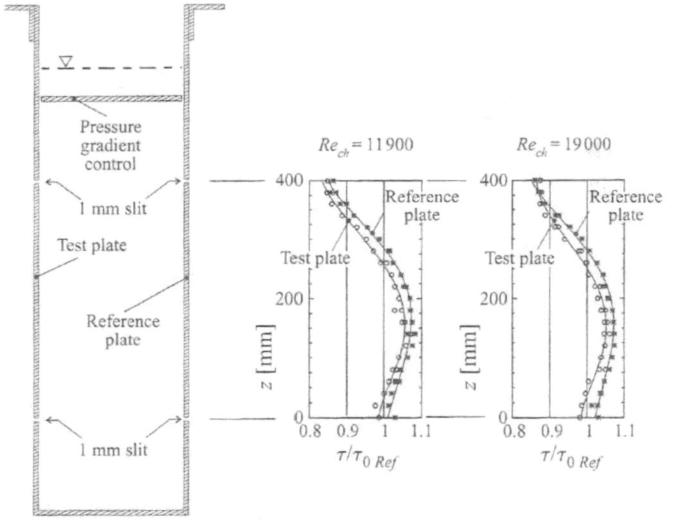

(b) Closed tunnel configuration

Figure 6.17: Cross sectional view of the test section of the Berlin oil tunnel and the corresponding shear stress distribution with open tunnel (a) and for zero pressure configuration (b) from Bechert *et al.* (1997). Reproduced with permission.

6.5 Conclusions

The investigation of riblet structures verified two important aspects. First the measurements are in very good agreement with the results of Bechert *et al.* (1997) which has been deemed to be the most accurate experimental data for riblets for the time being. A very good quantitative agreement between the results presented in this study and Bechert *et al.* (1997) is found: The optimum riblet spacing $s^+_{opt} = 16-17$ as well as 6% skin friction reduction match Bechert's data. For the present set-up the deviation from a uniform distribution of the skin friction can be recalculated exploiting equation 5.9 yielding a slightly higher drag reduction of 6.4%.

Second, the scatter of the data lies within the calculated uncertainty, supporting the careful and detailed uncertainty estimation in Chapter 4. Since the measurement error of $\tau_w/\tau_{w,0}$ can be assumed to be Gaussian distributed, the data scatter is a measure for the measurement accuracy. The direct comparison shows the accuracy of the present air flow facility to be comparable to the oil tunnel measurement from Bechert *et al.* (1992). The detection of the skin friction by using a pressure drop determination is thus comparable to the measurement with a high precision shear stress balance, having additionally the advantage of resolving the streamwise flow development.

For the studies of micro grooves as well as the dielectric elastomer actuators it was proven in this chapter that the set-up detects even smallest differences in skin friction.

An important result of this investigation, resolving previous uncertainties in the literature, is that the development of the flow over a riblet surface is not significant. However attention needs to be paid to the leading edge of the riblet plate. It is clearly shown that an abrupt onset of the structures induces disturbances which then cause a drag increase. This effect is certainly much more important for transitional flow investigations.

7 Micro grooves

Promising investigations on passive drag reduction by means of streamwise aligned grooves were conducted by Frohnapfel (2006) and Jovanović *et al.* (2010a). The authors supply a theoretical background and provide numerical as well as experimental data. In experimental studies significant drag reduction could be achieved: Frohnapfel *et al.* (2007a) reported a reduction of more than 20% in turbulent channel flows and Jovanović *et al.* (2011) quantified the drag reduction to be more than 30% in a transitional channel flow. The present investigation was intended to reproduce these results for transitional as well as for turbulent flow conditions utilizing the enhanced measurement accuracy of the present experimental setup in order to precisely determine the Reynolds number range, where drag reduction can be achieved. Furthermore, a quantification of the influencing parameters was planned.

The drag reducing capability of the micro grooves have only been verified in studies at LSTM in Erlangen. Other investigations on this type of small scale structure reduction are not known. Therefore, the introduction into the theory behind this flow control mechanism is primarily based on publications from Erlangen.

The fundamental concept of the effect of the micro grooves is based on statistical observations of the fluctuating velocity components in near wall turbulent flows. The directional anisotropy of the fluctuating velocity components of a turbulent flow is conveniently depicted in the so-called "anisotropy invariant map" (Lumley, 1977), which will serve as a primary element in the following considerations. Firstly, the theory and its application to drag reduction aspects are introduced following Frohnapfel *et al.* (2007b). In the following description the state of science is presented including investigations of transitional as well as turbulent flow conditions. Results of the present investigation are presented and the Chapter is completed with a discussion of the flow control effectiveness for both considered flow conditions.

7.1 Statistical consideration of turbulence via invariant map

The consideration of the character of a turbulent flow by means of invariants is based on the decomposition of a matrix into a symmetric and an asymmetric part. The procedure can be utilized to illustrate the statistical properties of turbulent fluctuations in the three directions of space. From the interpretation of the invariant map, conclusions about the state of the turbulence can be drawn. Especially the possibility to compare turbulent flows of differing surface topologies is valuable to understand the of manipulation of the shear stress in near wall flows.

In turbulence investigations the idealized concept of isotropic turbulence is often used for convenience. In this case the Reynolds stress tensor is not only symmetrical, but the components on its principle axis are equal $\overline{u_1 u_1} = \overline{u_2 u_2} = \overline{u_3 u_3}$. Especially in the near wall regions this assumption is not appropriate, since fluctuations of the component normal to the wall are dampened due to the no-penetration boundary condition. Thus the Reynolds stress tensor takes the form of an anisotropic tensor.

As any symmetrical tensor it can be split in an isotropic part $b_{ji}^{''}$ and an anisotropic part $b_{ji}^{'}$. The symmetric part notes $b_{ji}^{''} = \frac{1}{3} b_{kk} \delta_{ij}$, where δ_{ij} is the Kronecker-Delta and we can write the stress tensor:

$$b_{ij} = b_{ji} - \frac{1}{3} b_{kk} \delta_{ij} + \frac{1}{3} b_{kk} \delta_{ij} \tag{7.1}$$

The anisotropic part can be normalized using the trace of the tensor

$$a_{ij} = \frac{b_{ji}^{'}}{b_{kk}} = \frac{b_{ji}}{b_{kk}} - \frac{1}{3} \delta_{ij}. \tag{7.2}$$

Three scalar invariants can be computed from the normalized anistropic tensor (Jovanović, 2004):

$$a_{ij} = \frac{\overline{u_i u_j}}{q^2} - \frac{1}{3} \delta_{ij} \tag{7.3}$$

$$\mathrm{I}_a = a_{ii} = 0, \tag{7.4}$$

$$\mathrm{II}_a = a_{ij}a_{ji}, \tag{7.5}$$

$$\mathrm{III}_a = a_{ij}a_{jk}a_{ki}. \tag{7.6}$$

In the special case of axisymmetric turbulence the anisotropy tensor writes (Jovanović, 2004):

$$a = \begin{Bmatrix} b & 0 & 0 \\ 0 & -\frac{1}{2}b & 0 \\ 0 & 0 & -\frac{1}{2}b \end{Bmatrix}. \tag{7.7}$$

The scalar invariants are determined:

$$\mathrm{II}_a = \frac{3}{2}b^2, \mathrm{III}_a = \frac{3}{4}b^3. \tag{7.8}$$

Inserting the result for III_a in II_a yields:

$$\mathrm{II}_a = \frac{3}{2}\left(\frac{3}{4}|\mathrm{III}_a|\right). \tag{7.9}$$

Axisymmetric turbulence typically occurs when the flow is expanded or contracted, for instance, passing a nozzle or diffuser in a wind tunnel. If the sign of III_a is negative, it denotes an axisymmetric contraction, while the positive sign corresponds to axisymmetric expansion.

Considering the particular case of two-component turbulence the anisotropy tensor can be denoted:

$$a = \begin{Bmatrix} b & 0 & 0 \\ 0 & -b+\frac{1}{3} & 0 \\ 0 & 0 & -\frac{1}{3} \end{Bmatrix}. \tag{7.10}$$

The scalar invariants write:

$$\mathrm{II}_a = 2b^2 - \frac{2}{3}b + \frac{2}{9}, \mathrm{III}_a = b^2 - \frac{1}{3}b. \tag{7.11}$$

The relation between invariants is:

$$\mathrm{II}_a = \frac{2}{9} + 2\mathrm{III}_a. \tag{7.12}$$

If II_a is plotted versus III_a, the anisotropy invariant map is derived (Fig. 7.1), wherein the terms for axisymmetric (Eq. (7.9)) and two-component turbulence (eq. (7.12)) form the boundary lines of a curvilinear triangle. According to Lumley (1977) all realistic states of turbulence are contained within this triangle. The intersection points of the boundary lines correspond to isotropic turbulence ($\mathrm{II}_a = \mathrm{III}_a = 0$), isotropic two-component and one-component turbulence.

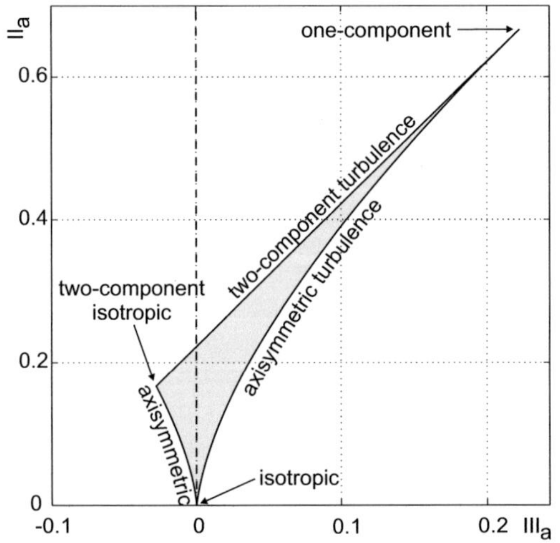

Figure 7.1: Anisotropy invariant map.

7.1.1 Examination of skin friction reduction with respect to the anisotropy of turbulence

In Chapter 2 it has been shown that for a turbulent channel flow the total shear stress (Eq. (2.31)) results from two sources: the viscous shear stress

$\eta \mathrm{d}U_1/\mathrm{d}x_2$ and the Reynolds stresses $\rho\overline{u_1 u_2}$. If the turbulent channel flow is considered from an energetic standpoint, these stresses cause dissipative losses, which lead to the pressure drop in streamwise direction. Thus, the required power P to drive the flow through the channel, can also be expressed in terms of the average energy dissipation rate per unit fluid mass $\bar{\tilde{\epsilon}}$

$$P = \Delta p\dot{V} = \bar{\tilde{\epsilon}}m, \text{ with } m = \rho HWL. \tag{7.13}$$

The average dissipation rate $\bar{\tilde{\epsilon}}$ is given by the integration of the total energy dissipation rate $\tilde{\epsilon}$ over the volume of the channel

$$\bar{\tilde{\epsilon}} = \int_V \tilde{\epsilon} dV, \tag{7.14}$$

where $\tilde{\epsilon}$ splits in the two terms of the direct and the turbulent dissipation

$$\tilde{\epsilon} = \nu\left(\frac{\partial\overline{U_i}}{\partial x_j} + \frac{\partial\overline{U_j}}{\partial x_i}\right)\frac{\partial\overline{U_j}}{\partial x_i} + \nu\overline{\left(\frac{\partial u_i}{\partial x_j} + \frac{\partial u_j}{\partial x_i}\right)\frac{\partial u_j}{\partial x_i}}. \tag{7.15}$$

Considering DNS data Frohnapfel *et al.* (2007c) show the turbulent dissipation to be dominating with increasing Re, which are found in application. In consequence, to obtain large drag reduction in application the turbulent dissipation needs to be reduced.

In numerical investigations by Lammers *et al.* (2006) and Frohnapfel *et al.* (2007c) the relation between the turbulent dissipation and the state of high anisotropy of fluctuation components is examined. Frohnapfel *et al.* (2007c) investigate the effect of an artificially generated one-component state of the near wall turbulent fluctuations on the shear stress. Directly at the wall the flow is forced to an axisymmetric flow state by imposing spanwise velocity fluctuations. These artificial limitations are extended stepwise to the complete range of the viscous sublayer. Frohnapfel *et al.* (2007c) found a decrease of the turbulent dissipation in the near wall region, resulting in a lower wall shear stress. With growing area of the artificial boundary condition, τ_w is successively lowered. These results demonstrate the relation between the state of anisotropy of the turbulence and the amount of dissipated energy in a wall bounded turbulent flow. However, this investigation does not propose a mechanism to achieve the one-component state of turbulence in the near wall region in a real application.

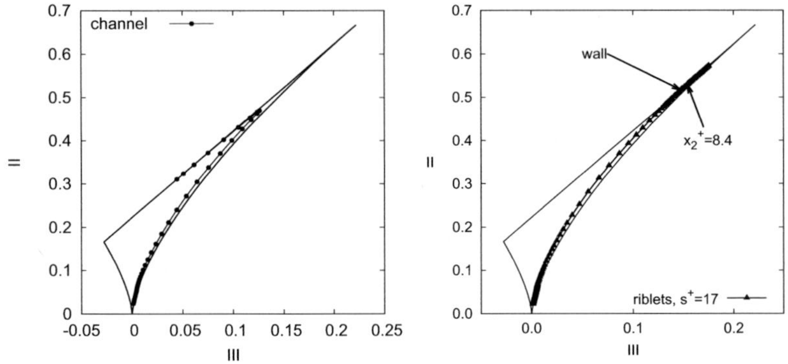

Figure 7.2: Trajectory from channel the centerline to the wall: For smooth surface (left), to the valley inbetween riblets (right). The arrows in the right picture indicate the height of the riblet tips and the end point at the wall.

Subsequent Frohnapfel (2006) by means of the anisotropy consideration analyzed numerical data from a number of flow control techniques which effectively manipulated the friction drag. Riblets, long-chain polymers and surfactant additives as well as blowing and suction were regarded. Further flow control by means of magnetic fields, laminarisation of severe accelerated flows and compressibility effects leading to decrease of c_f for increasing Mach number were analyzed. In all drag reducing cases a coinciding trend towards a state of highly anisotropic turbulence near the wall is found. It is concluded that in order to gain a high amount of drag reduction, near wall velocity fluctuations need to be predominantly one-componental. Based on this idea of "a common mechanism for drag reduction" (Frohnapfel, 2006), the anisotropy invariant map was found to be an effective tool to identify a basic mechanism of turbulent drag reduction.

To demonstrate the interpretation, the invariant map of riblet data obtained in a DNS (Fink, 2014) is exemplarily shown in Figure 7.2. The plots show the anisotropy development in a channel from the channel center to the wall as trajectories in the invariant map. The left picture depicts the result for a smooth channel. In the channel center nearly isotropic turbulence is obtained, while at the wall the state is two-componental. In the right diagram the result from the channel center to the valley between the riblet tips is shown. The trajectory tends significantly more to the one-component state in the near-wall region. If we compare this observation to the functional principle of riblets (see Chap. 6), the tendency to one-component turbulence is found in the region of the lower skin friction.

These findings deduced from statistic examination of turbulence are exploited for an approach to reduction of skin-friction drag by (Frohnapfel, 2006), which is introduced in the following Section.

7.2 Micro grooves - State-of-the-art

7.2.1 Basic idea of micro grooves

The basic idea of micro grooves is to generate a region of one-component turbulence at the wall. If we consider the near wall region, the velocity fluctuation component normal to the wall u_2 is vanishing. A two-component state of the turbulence is reached. Imagine the surface to be equipped with streamwise orientated grooves, which have a rectangular cross section. Inside the grooves the velocity component in x_2-direction is again dampened through the bottom wall of the groove. In the spanwise direction fluctuations are hindered by the side walls, if the width a is small. In consequence the turbulence within the small grooves must be one-componental. In terms of the invariant map the flow inside the grooves needs to simultaneously fulfill the criteria of two-componentality and axisymmetry. This state is represented by the upper edge of the triangle which is corresponding to the maximum anisotropy. As a result the fraction of the surface area covered by the grooves exhibits a state of high anisotropy. This is contrary to a smooth surface, where in the near-wall region a two-component turbulence state is obtained on the entire surface. A prediction how the partly one-component state on a grooved surface will influence the skin-friction is described by the "hydraulic model" which has been introduced by Frohnapfel *et al.* (2008).

The hydraulic model describes a simplifying method to predict the drag reduction potential of a surface with streamwise grooves. Figure 7.3 illustrates the fundamentals of the model. For a channel the friction coefficient of turbulent flow according to Dean and a laminar flow $12/Re_b$ are shown over Re_b. It is assumed that the laminar flow will persist up to high Reynolds numbers within the grooves. Regarding the c_f-values for increasing Re_b,

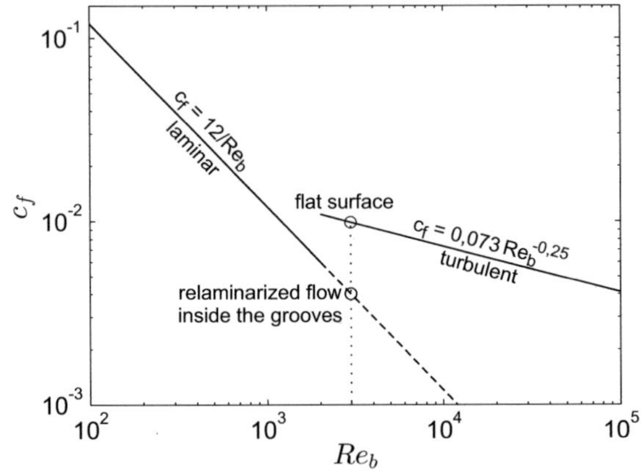

Figure 7.3: Comparison of the laminar and the turbulent friction coefficient in a channel flow.

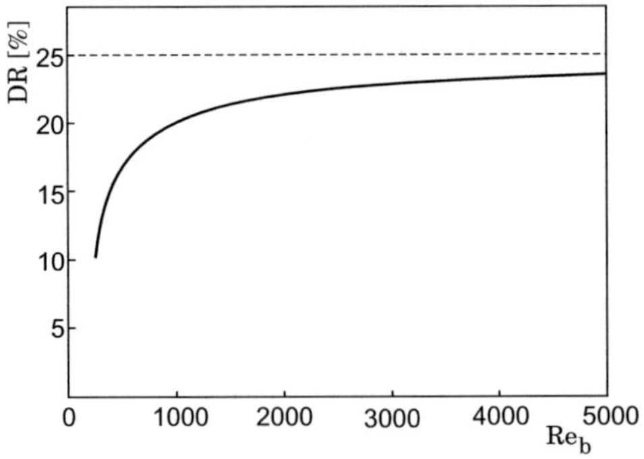

Figure 7.4: Drag reduction DR predicted by the hydraulic model, adopted from Frohnapfel (2006).

the laminar value is much smaller than the corresponding turbulent value. Based on the assumption that the skin friction within the grooves is negligible compared to the skin friction on the smooth plateaus inbetween the grooves of large Reynolds numbers, $\tau_w/\tau_{w,0}$ will decrease by approximately the percentage of the surface being covered by the grooves.

The hydraulic model yields the prediction for the drag reduction shown in Figure 7.4, where the drag reduction DR is defined as

$$DR = 1 - \tau_w/\tau_{w,0}. \tag{7.16}$$

The model assumes laminar flow on the complete surface inside the grooves and uses the Dean's law for the surface area assumed to be covered by turbulent flow. Thereby the additional friction loss originating from the upper edges of the grooves are neglected. The model estimates the drag reduction, but it cannot predict the scaling of the grooves. To verify the model Frohnapfel (2006) compared DNS simulations of a channel flow with smooth walls and a channel equipped with streamwise grooves. The dimensions in width and depth of the grooves were chosen about the size of the viscous sublayer of the flow on a smooth wall. The simulations were conducted with an equidistant grid (cf. Frohnapfel 2007). Thus a grid refinement for a high resolution of the flow in the region of the grooves was not used. Three simulations were conducted for centerline Reynolds numbers of Re_c=5230, 6650 and 6690, each for slightly different groove geometries. A fraction of 25% to 33% of the surface was grooved. The results showed very good agreement with the predictions of the hydraulic model for the wall shear stress, but the desired reduction of the turbulent dissipation was not found for the selected scaling. In contrast additional fluxes originating from secondary motions were found to outbalance the gain obtained for the direct dissipation.

7.2.2 Previous results for grooved surfaces in turbulent flows

For a parameter study Frohnapfel *et al.* (2007a) also conducted experiments in a channel flow facility. The used wind tunnel (see Spudat (1981)) is of the same type as the one developed in the present work and the experimental set-up and procedure are very similar. The width of the channel is the same and the length is only half a meter shorter with 3.5m. In contrast to the present investigation the tunnel can be equipped with different side walls of H =25, 35 and 41mm corresponding to AS = 12, 8.6 and 7.3. The flow was tripped with a similar 12% tripping and the turbulence level was measured with a hot-wire installed at the outlet of the tunnel.

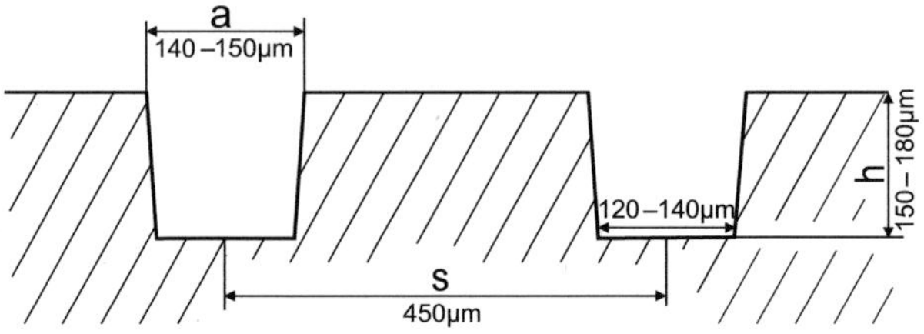

Figure 7.5: Geometry of the structures used in Frohnapfel *et al.* (2007a).

For the experiments a channel configuration with a smooth development section of half a meter was followed by a section with structured surface and a smooth reference surface, each 1.5m long. On the bottom and top wall the structured section was equipped with rectangular grooves with a width of $a = 140 - 150\mu$m and a depth of $h = 150 - 180\mu$m. The groove spacing was chosen to $s = 450\mu$m (Fig. 7.5). The pressure loss was measured along the channel to determine the friction losses. The skin friction reduction was measured by a similar strategy as presented in Chapter 6. Frohnapfel *et al.* (2007a) defined the drag reduction DR using the ratio of pressure loss per running length $\Delta p/\Delta L$ on the grooved compared to the smooth surface:

$$DR = 1 - \frac{(\Delta p/\Delta L)}{(\Delta p_0/\Delta L_0)}. \tag{7.17}$$

Figure 7.6 shows the results for the 25mm and the 35mm test section (Frohnapfel, 2006). While in a narrow range for quite small Reynolds numbers a large DR was indicated, the grooves obviously did not influence the skin-friction on the plates at other Re_b. The scaling of the grooves was normalized as

$$a^+ = \frac{a\, u_\tau}{\nu} \tag{7.18}$$

with

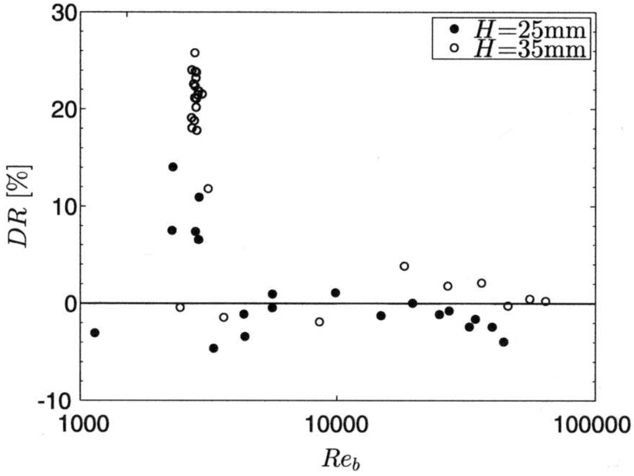

Figure 7.6: DR in dependence of Re_b for channel height of 35mm (empty symbols) and 25mm (filled symbols) adapted from (Frohnapfel, 2006).

$$u_\tau = \sqrt{\frac{\Delta p_0}{\Delta L_0} \frac{WH}{2\rho(W+H)}}. \tag{7.19}$$

The expression takes into consideration the presence of the sidewalls. For large AS it is equivalent to the normalization used in Chapter 6, where the skin friction is calculated with the assumption of a two-dimensional flow. The maximum DR of the grooves corresponds to $a^+ \approx 1$ for both channel heights. On the one hand the obtained scaling is significantly smaller than the dimensions focused in the numerical results. On the other hand Grüneberger & Hage (2011) showed that very small scale riblet structures in the order of one wall unit can have considerable influence on the skin friction in a turbulent flow. A drag reduction of about 0.5% was obtained even for the small scale riblet structures, although being considerably less than the numbers indicated in the previous micro groove investigations.

A hint to the question of the required scaling of the grooves is given in a riblet study by Walsh (1982). Walsh's intention was to study whether the drag reduction effect of riblets is based on the sharp protruding tips or on the grooves inbetween. Therefore two semicircular shaped test plates were produced, one with convex and one concave structuring (see Figure 7.7 (a)). The convex structure is very similar to the optimized geometry for the micro grooves suggested in (Frohnapfel *et al.*, 2007a). The result of $\tau_w/\tau_{w,0}$ for both structure types is shown in Figure 7.7 (b). While the concave curvature decreased the skin friction by 7% at a spacing of approximately $s^+ = 15$, the convex structure indicated no influence for the same spacing. At a spacing of $s^+ = 32$ the concave structure turned to a drag increase, while the convex structure did not affect the friction drag for $s^+ < 40$. The drag increase at the same spacing was much lower than for the concave structure. Additionally, V-groove structures with a flat interspace have been investigated with a similar result (Walsh, 1982). Walsh's investigation shows that at quite large spacings, up to $s^+ < 40$, structures similar to the micro grooves do not influence the drag. This supports the finding by Frohnapfel *et al.* (2007a) that the micro grooves do not increase the skin friction drag even at quite large Reynolds numbers. In Walsh's study only the concave riblet like structuring showed drag reduction. The convex surface could only increase the skin friction. However, the study by Walsh (1982) does not show results for $s^+ < 10$, where the drag reduction of the micro grooves was found for $s^+ \approx 3$.

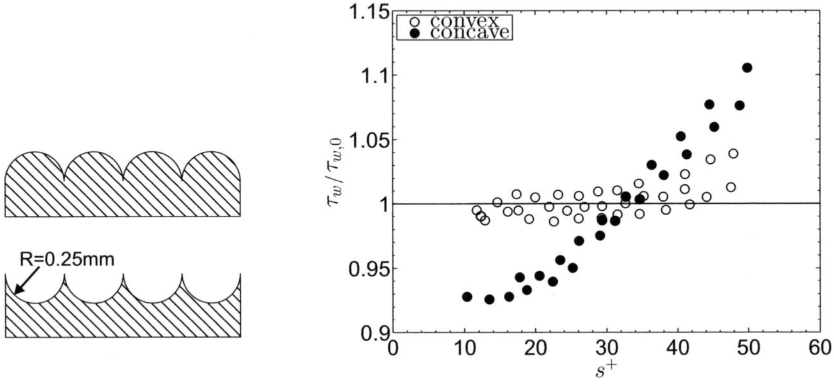

(a) (b)

Figure 7.7: Dimensions of semi-circular grooves studied by (Walsh, 1982) (a). $\tau_w/\tau_{w,0}$ in dependence of s^+ for convex and concave surface structuring (b) adopted from (Walsh, 1982).

7.2.3 Previous results for grooved surfaces in transitional flows

Since the mechanism to laminarize the flow inside the grooves was assumed to work in the same manner in a transitional flow, an experimental study in naturally developing flow was conducted by Jovanović *et al.* (2011) using the same experimental set-up. The arrangement with a smooth development section of 1m, a structured section of 1.5m and a smooth downstream section of 1.5m was identical to the configuration used for the turbulent investigation. The grooved plates were the same as in the turbulent flow investigation, but the tripping device has been removed to enable a natural development of the flow. The measurement strategy needed to be changed from the direct comparison of two subsequent sections in simultaneous measurement to the comparison of two separate measurements, one of a the partly structured channel and the other of the completely smooth reference test section.

The comparison of the pressure drop measurements was presented in two different ways. First the total pressure drop over the downstream section of the channel was examined and second a comparison of the development along the channel is investigated.

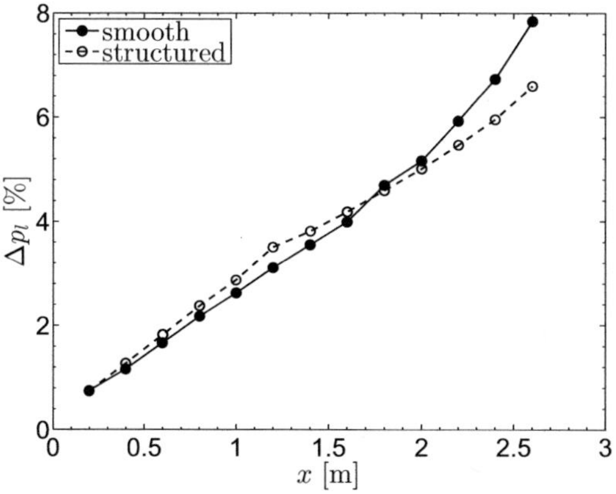

Figure 7.8: Comparison of the pressure drop in a completely smooth channel and a channel with a structured section for the highest DR at $Re_b = 6930$ obtained by Jovanović *et al.* (2011).

Figure 7.8 (adopted from Jovanović *et al.* (2011)) shows the pressure drop along the channel for a smooth and a partly structured surface condition at $Re_b = 6930$. Note that the length of the development section plus the distance to the first pressure tap, in sum 0.7m, is not included in the data. Thus data presentation starts directly on the structured plate. The pressure drop on the grooved surface is slightly larger than on the structured plate for $x < 1.5$m. While the pressure follows a laminar development almost to the end of the channel downstream the grooved surface, it increases in the downstream section of the completely smooth channel. The plot depicts the situation for the highest DR found at $Re_b = 6930$. For a narrow range of Re_b a delay of the transition is obtained. For smaller and larger Reynolds numbers the grooves have no influence on the flow, which is in agreement to the findings in turbulent flow.

As a result, the partly structured channel exhibits a smaller pressure loss, but drag reduction is not found, where the grooves are located. In contrast, the skin friction is little higher on the structured section. Jovanović *et al.* (2011) explain the increase of pressure drop in the structured part of the test section as the interaction of the microgroove structures with near-wall disturbances. This would lead to a stabilizing effect and prevent the growth of disturbances in the near wall region. Therefore the flow would develop laminar over the downstream smooth section.

Furthermore, Jovanović *et al.* (2011) presented a comparison of the turbulence level close to the channel outlet. Separate plots for the grooved and the smooth surface were shown, each for increasing and decreasing flow rate. A significant hysteresis effect between the increasing and decreasing case is observed for both surface conditions. It is concluded that for the verification of a transition delay the shift of the transition to higher Re needs to be achieved for identical inlet conditions with increasing flow rate (Jovanović *et al.*, 2011).

To allow a direct comparison of the data for increasing flow rate, the present author has adopted the data of both surface conditions from Jovanović *et al.* (2011), which is shown in Figure 7.9. The laminar-turbulent transition is clearly visible in each curve. The position of maximum turbulence intensity is nearly equal for both surface conditions. Only for low Reynolds numbers the results differ slightly. The turbulence intensity grows continuously for the smooth surface condition, whereas it rather stagnates before rising abruptly at $Re_b = 6000$ for the grooved surface condition. In the fully turbulent regime, the curves almost coincide. Comparing these two curves

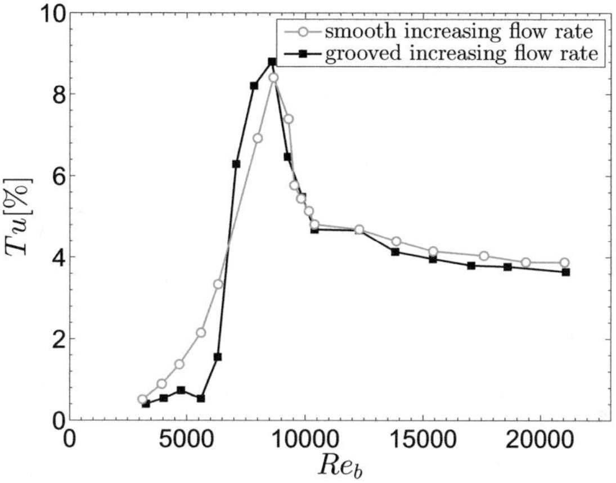

Figure 7.9: Comparison of the turbulnce level for increasing flow rate for a complete smooth channel and a channel with a structured section (adopted from Jovanović *et al.* (2011)).

in the fully turbulent part, Jovanović *et al.* (2011) illustrate a shift of the state of fully developed turbulent flow to higher Re in the grooved channel. This interpretation of the data is not shared by the author as a significant shift is not visible in Figure 7.9. For $Re_b > 9000$ the observed turbulence level is the same for both surface conditions.

7.3 Experimental investigation of micro grooves

The literature data indicated the effective groove width in turbulent flow at very small scaling of $a^+ = 0.8$ with corresponding Reynolds number being close to a still transitional regime (cf. (Frohnapfel *et al.*, 2007a)). Thus the goal is to produce slightly smaller structures to guarantee a fully turbulent flow conducting the experiments at higher Re. Identical grooves have shown to be effective in transitional and turbulent flow for significantly different Re_b, resulting in a different scaling in wall units. Therefore a small difference in the groove size is assumed not to change the effectiveness.

The rectangular grooves are shown to reduce skin friction drag in Frohnapfel (2006), although the geometry exhibits two friction generating hot-spots at the edges. Further improvement of the effectiveness was intended by a smaller angle of the groove flank with using V-shaped geometry.

7.3.1 Fabrication of the groove structures

The fabrication of the grooves turned out to be quite difficult. The challenging task was to produce the geometry on such a large surface area within the specified accuracy. Most micro-machining applications and in consequence the machine tools are also of small size. Precise manufacturing is usually limited to few centimeters with an enormous time effort. For example, micro-milling produced very good results, but to structure a plate of 1500mmx300mm would have taken two months.

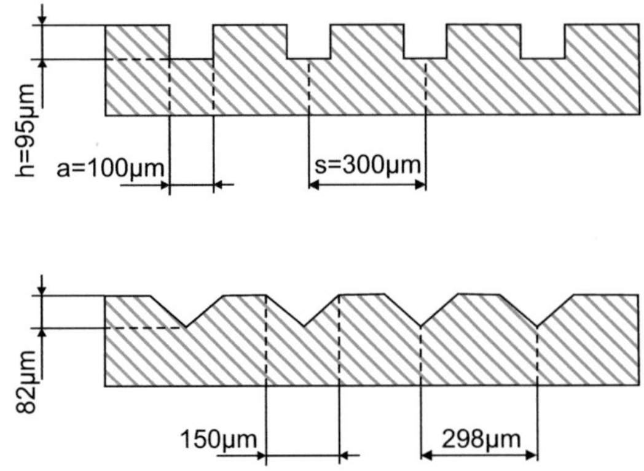

Figure 7.10: Dimensions of rectangular and V-shaped grooves.

To resolve these issues, the structures have been successfully manufactured by broaching aluminum plates. The fabrication has been conducted at the Institute PTW at TU Darmstadt. It has been possible to produce three segments of 500mm length. A conventional milling machine was equipped with a custom built broaching tool, which allowed to simultaneously broach 18 parallel grooves. The best results were obtained for plates of high-strength aluminum. The surface needed to be machined two times to achieve the required depth of the grooves with a high-quality surface finish inside the grooves. Two different grooves types were realized: a triangular and a square groove shape. After the broaching process significant burrs occurred along the upper edges of the grooves for the rectangular geometry. These plates were sanded until the burrs had been removed without a detectable removal on the smooth surface between the grooves. The final geometry of both groove geometries is shown in 7.10. In both cases the geometry is very precise. The deviation of the groove depth and spacing have been checked with a microscope to have a tolerance of only $\pm 2\mu$m. The sidewalls of the rectangular grooves are nearly normal to the horizontal bottom surface (89.6°).

7.3.2 Measurement strategy

For the transitional as well as for the turbulent study measurements in the channel with a partly structured surface were compared to the corresponding complete smooth reference data, which are shown in Chapter 5. In both measurement campaigns the humidity has not been taken into account. Therefore measurement accuracy is decreased in comparison to the riblet study. A measurement accuracy of changes in $\tau_w/\tau_{w,0}$ of 2% for $Re_b < 4000$ increasing to slightly less than 1% at $Re_b \geq 8000$ is possible. Considering the expected large amount of $DR = 10 - 20\%$ on the structured surface, this is still appropriate.

7.4 Measurements in turbulent flow

For the measurements in turbulent flow the surfaces structured with rectangular grooves were installed in the downstream part of the channel to guarantee a fully turbulent flow at preferably low Re. The 13% tripping has been used.

Figure 7.11 compares the pressure drop of the reference measurement with smooth surface to the results obtained with a grooved section in the downstream part of the channel. In total the plots consists of 450 data points. Although it is not visible in this presentation due to the very good matching, the filled symbols at higher Re_b are not single measurements, but were repeated between four to six times. The entire Re range was checked in very smalls steps of the flow rate, but since no difference was obtained, measurements were concentrated more in promising low Re range.

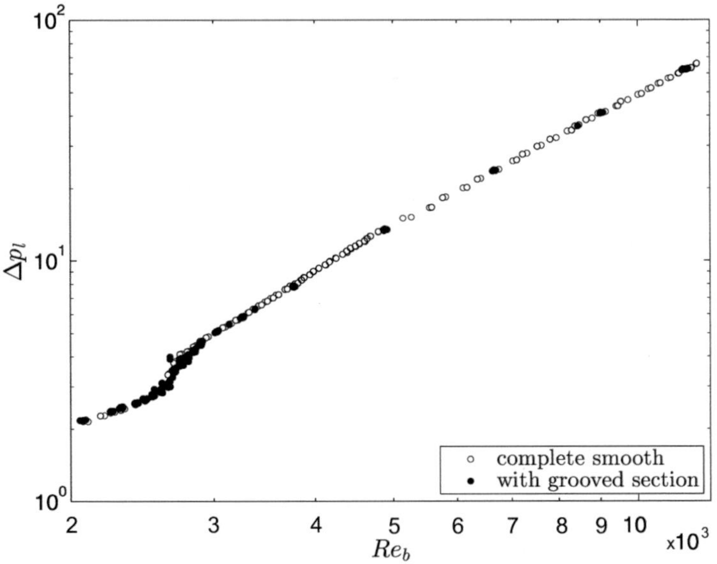

Figure 7.11: Comparison of the pressure loss in a complete smooth channel and a channel with a structured section in dependence of Re_b.

From the evolution of the pressure drop results the range of laminar ($Re_b < 2500$), transitional and turbulent flow ($Re > 4000$) can be differentiated. For the laminar as well as for the turbulent regime the data show the same result for both surface conditions. Only in the highly transitional state, for $2600 < Re_b < 3000$ a small differences between smooth and structured surface can be conjectured.

For the transitional regime it was mentioned in Chapter 5.3, that the environmental noise had an influence on the pressure loss (up to 3%). Therefore, the measurements in the transitional region were completely repeated under absolute quiet conditions for the grooved as well as for the reference surface. The test section has been reassembled changing between the two surface conditions. Figure 7.12 focuses on the measurements in the transitional region. The data show the c_f-value instead of the pressure loss to exclude slight deviation between different measurements originating from the changes in the air properties. The high reproducibility confirms the estimated measurement uncertainties of approximately 2% at low Re and is in the range of 1% for Re_b=3500. In the transitional regime a naturally

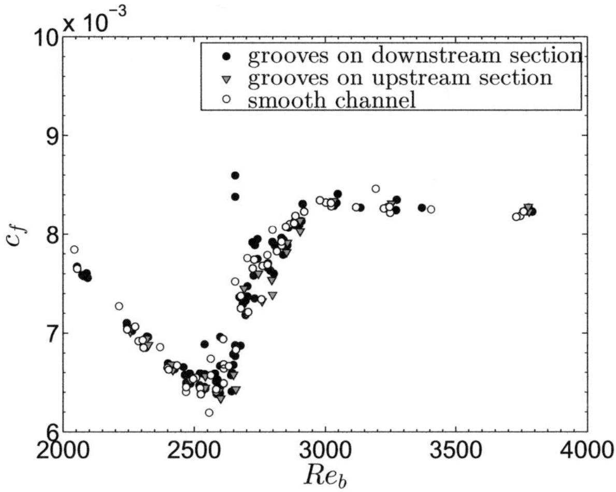

Figure 7.12: Comparison of c_f in the completely smooth channel and the channel with a structured downstream section and the channel with a structured upstream section.

larger scatter is observed. Again no difference is observed for the skin friction coefficient between the two surface conditions. To investigate a potential effect of the grooves on the transition, the grooved plates have been installed in the upstream section of the channel and the measurement was repeated. Figure 7.12 also includes these data. However, the obtained conclusions are the same as in the case with grooves on the downstream section.

7.4.1 Discussion of results in turbulent channel flow

Since the measurement results did not indicate an effect of the grooves, even at high *Re*, the procedure has been rechecked, but the measurements were confirmed to be correct. A special focus on the transitional regime was dedicated to the fact, that in the previous investigation of Frohnapfel *et al.* (2007a) drag reduction was obtained very close to the transitional region. From the hot-wire measurements of Frohnapfel (2006) it can be construed that the flow was not fully turbulent for $Re_b = 3000$. The author initially expected the effect of the grooves to be related to a transition delay.

Although a delay of transition could not be observed, the explanation for the DR in the former turbulent investigations may be related to a not fully turbulent flow.

A consideration of the flow development highlights a possible explanation for the discrepancy between the similar investigations in different test facilities: Even if Re_b is slightly above the fully turbulent limit at the channel outlet, a considerable length of laminar to transitional flow in the tunnel entrance region can be conjectured. In the present facility this was demonstrated in the reference measurement in Chapter 5 and it is shown explicitly in Figure 5.9(b), where most of the channel length shows a laminar flow development. The observation remained similar also for the 16% tripping, where the flow at the outlet was already fully turbulent at the smallest considered Re (compare Fig. 5.10).

In Figure 7.13 the original data from Frohnapfel *et al.* (2007a) are plotted on the left hand side and the data from the smooth reference of the present study on the right hand side. Since the development length of 700mm from the channel entrance was not studied in the Frohnapfel's investigation and the present channel is 0.5m longer, these additionally investigated sections in the present investigation are shaded in grey. The lines indicate the mean pressure gradient on the smooth and the structured segments. The slightly deviating tripping device and the differing channel height (35mm and 25mm) lead to differences in the flow development at the same Re_b between the two experimental set-ups. Additionally, the crucial differences in the determination of Re_b induce considerable uncertainty to the comparability of the Reynolds number (cf. Appendix A.1). While even at very low flow rates the present instrumentation has a measurement accuracy better than $\pm 1.5\%$ for Re_b, the uncertainty of Re_b is specified with $\pm 10\%$ in Frohnapfel (2006) for very low Re. Thus corresponding states of flow development are compared. Note that the diagrams on the left-hand side do not depict the physical pressure loss but the voltage output of the pressure transducer at varying measurement ranges of the sensor as given by Frohnapfel *et al.* (2007a).

The comparison of the larger Reynolds numbers ($Re_b = 3620$ and 3940, respectively) shows a more or less constant pressure drop in the considered part of the test section. The diagram (f) for $Re_b = 3940$ indicates a laminar region at the beginning of the channel (within the shaded section). For $Re_b = 3300$ (d) the laminar region propagates far into the channel resulting in a smaller gradient in the downstream section than in the upstream

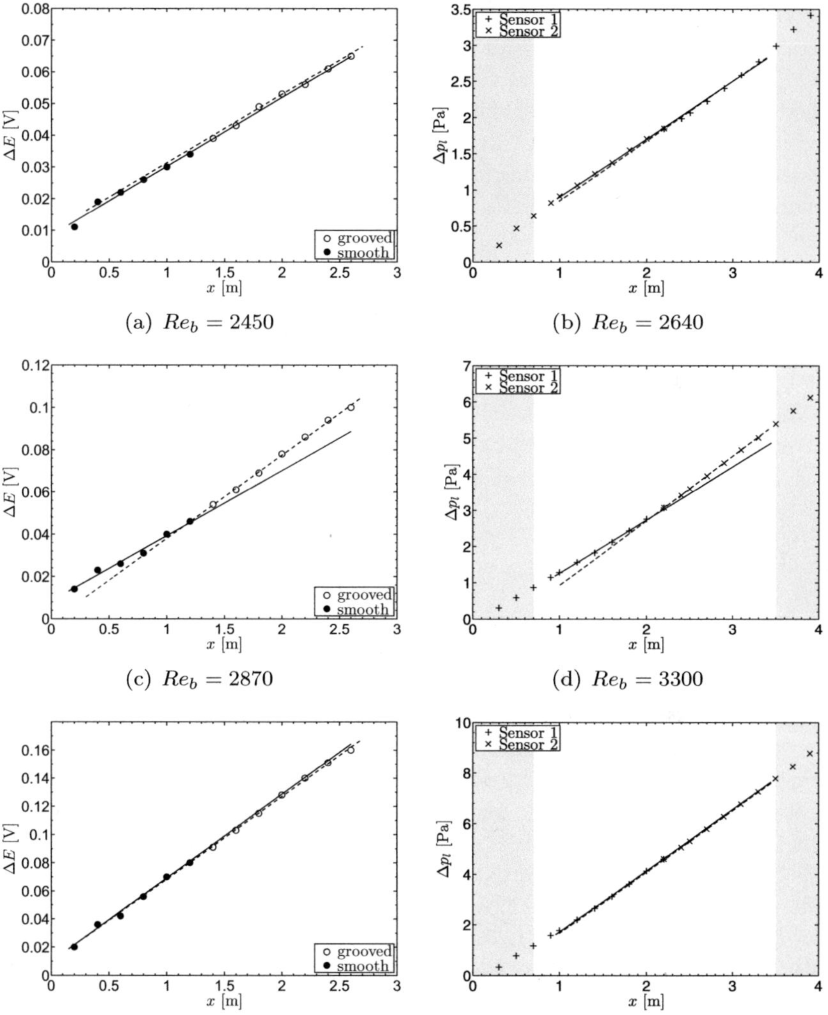

(a) $Re_b = 2450$ (b) $Re_b = 2640$

(c) $Re_b = 2870$ (d) $Re_b = 3300$

(e) $Re_b = 3620$ (f) $Re_b = 3940$

Figure 7.13: Comparison of the pressure drop for the complete test sections. Data with the upstream section equipped with grooves are replotted from Frohnapfel *et al.* (2007a) in the left column. The smooth reference of the present measurement are shown in the right column for different Re_b. Dashed lines depict the mean pressure gradient on upstream and continuous lines on the downstream section.

section. This finding is very similar to $Re_b = 2870$ (c), which depicts the data for the highest DR from Frohnapfel *et al.* (2007a). For even lower Re_b, Figures 7.13 (b) depicts the flow development for a Reynolds number, where the hot-wire measurement at the channel outlet still indicated a high turbulence level. Although the flow is transitional for most of the channel length, the pressure gradient along the channel is approximately constant, if the sections shaded in grey are not taken into account. Figures 7.13 (a) shows the corresponding data from Frohnapfel *et al.* (2007a).

Determining the drag reduction by an evalution of the ratio of the mean pressure gradient along the upstream and downstream plate as done by Frohnapfel *et al.* (2007a), a significant drag reduction would be construed for both situation in Figures 7.13 (c) and (d). However, it can be clearly seen from the present data that the different gradients are exclusively a result of the flow development in the channel. This also can explain why Frohnapfel only obtained drag reduction in the case of grooves in the upstream section. The observation of a smaller gradient in the downstream part is also observed for a completely smooth channel by Frohnapfel *et al.* (2007a) at $Re_b = 2850$, which is very close to $Re_b = 2870$ of maximum DR with the structured surfaces (compare Fig. 7.13 (c)).

The increase of the drag reduction for different channel heights can be interpreted as follows. Note that at the same flow rate channels of arbitrary height exhibit the same Reynolds number (for a constant channel width W), thus $Re_b \neq f(H)$. Yet, the flow development is a function of the channel height, meaning that with increasing channel height, the required physical development length is longer for a given Re_b. Hence, the transitional length, which is found at the channel entrance for low Re even with the tripped flow, can probably extend further into the channel for larger H and thus the apparent DR is increased. These considerations can explain the increased DR from a $H = 25$mm to $H = 35$mm in Frohnapfel (2006) (compare Fig. 7.6). The same tendency can be construed from the significantly larger DR reported by Jovanović *et al.* (2010a) with a channel configuration using $H = 41$mm. As soon as the Reynolds number is large enough to enable an abrupt turn to a turbulent state, a running length of 700mm will be sufficient to obtain a fully turbulent flow. Due to the required development length, which was found to be in range of $H = 70 - 80$ a comparison of the pressure gradient on the downstream and upstream section will show only slightly smaller values for the downstream section. This is indicated in a comparison of pressure drop on the two section of a complete smooth channel shown by Frohnapfel *et al.* (2007a) for higher Reynolds numbers.

In summary, the drag reduction due to micro grooves, as reported by of Frohnapfel *et al.* (2007a) could not be confirmed by the numerous more sophisticated and more accurate measurements. A significant effect of the groove structures on turbulent channel flow could not be verified.

7.5 Measurements in transitional flow

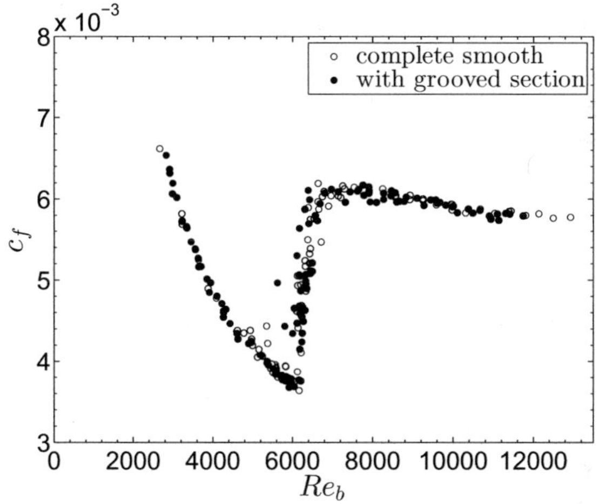

Figure 7.14: c_f over Re_b for complete smooth and partly grooved surface for naturally developing flow.

In the previous measurements the rectangular grooves were shown to be practically effectless in turbulent flow. In general, the V-shaped grooves were supposed to be more effective, since the friction generating upper edges at the groove flanks are reduced in angle. To maximize any possible drag reduction effect in transitional flow, the V-shaped geometry is used for this more difficult investigation in which the flow is developing completely naturally, i.e. no tripping is used. The channel configuration is similar to the one presented by Jovanović *et al.* (2010a): A smooth entrance section of 950mm length is followed by a 1500mm segment with the structured surface and the downstream section of 1500mm is again smooth. The difference to the measurement by (Jovanović *et al.*, 2010a) is the length of the smooth

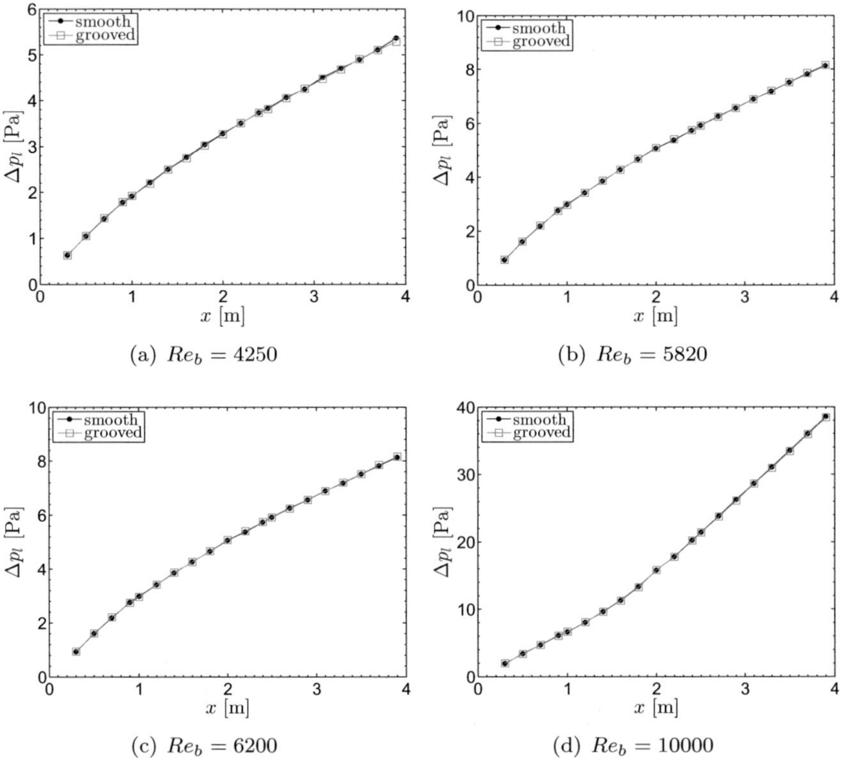

(a) $Re_b = 4250$

(b) $Re_b = 5820$

(c) $Re_b = 6200$

(d) $Re_b = 10000$

Figure 7.15: Comparison of the pressure drop for the completely smooth test section and the one with a grooved section for different Re_b.

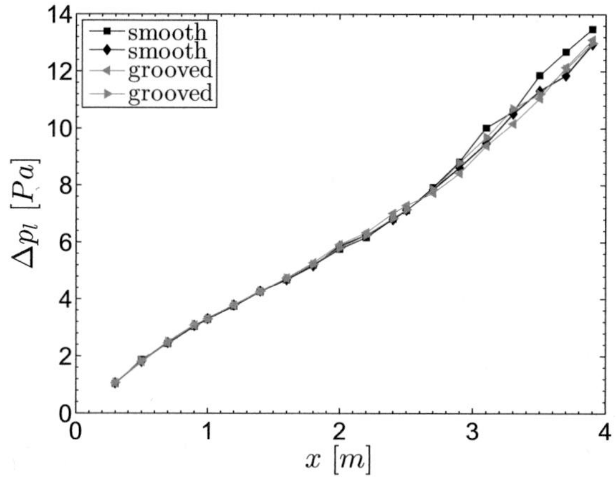

Figure 7.16: Pressure drop for the transitional regime at $Re_b = 6320$ for different measurements.

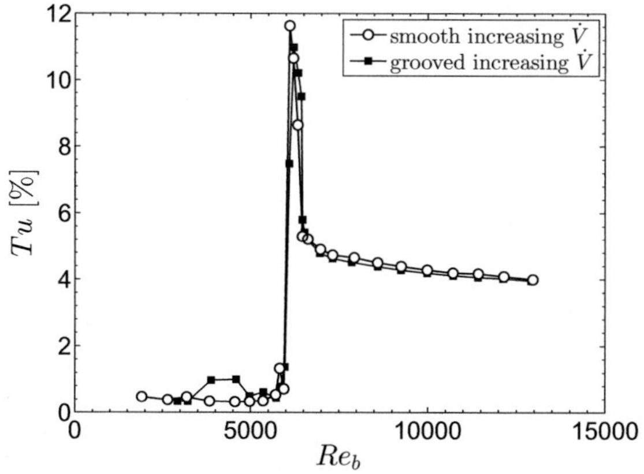

Figure 7.17: Turbulence level for increasing and decreasing Re_b.

section at the channel entrance, which is 450mm longer than the 500m section used by Jovanović *et al.* (2010a).

Figure 7.14 shows the skin friction coefficient over Re_b for the completely smooth channel and the channel configuration with a structured section. Strikingly, very similar results are obtained for both surface conditions for all Reynolds numbers. In the range of $5000 < Re_b < 6000$ few scattering data points indicate how slight environmental noise can trigger premature transition, leading to an increase of the skin friction. Generally, scatter in the data is larger in transitional range. It decreases significantly for $Re_b > 9000$, the Reynolds number range where the flow directly develops turbulent.

To identify possible differences in the streamwise flow development, Figure 7.15 presents the pressure drop along the channel with a grooved section in comparison to the data of the smooth reference channel presented in Chapter 5.1 for different *Re*. In all cases a nearly perfect agreement of data of the structured and smooth plates is observed. In the transitional region more scatter of the data was found, since very small differences in the *Re* cause large differences in the pressure drop. Even the temporal evolution plays a role. For a narrow Reynolds number range ($6100 < Re_b < 6700$), it was found that the transition can vary slightly towards the turbulent or towards the laminar flow state during long-time measurements. However, this effect could only be observed in a narrow range of *Re*. Figure 7.16 examplarily shows four measurements for the *Re* at which very high turbulence level at the channel outlet was obtained. Two measurements with completely smooth surfaces and two measurements with partly structured channel walls are compared for the same Reynolds number ($Re_b = 6320$). At the channel entrance the development of all four measurements is very similar. At the half length of the channel there are small deviations, although the total pressure loss is almost the same in all cases. Even for this highly transitional state, significant differences in the flow development depending on the surface conditions are not observed.

To exclude an influence of an improper adjustment of the test plates during the assembly, the channel configuration has continuously been changed between the smooth and structured surface. The presented measurements are part of the data gathered over a period of four months. Note that there is no indication of an influence of the test plate intersections in any of the data sets. As mentioned in Chapter 5.1, the pressure tab at 2m did not work properly, but is shown since the deviation is small and reproducible.

The measurement of the turbulence level showed the same result for both surface conditions for increasing flow rate (see Fig. 7.17). For decreasing flow rate the result differs significantly between the measurements. It seems that even for very calm conditions the turbulence level can decrease rapidly with only little difference to the increasing flow rate. In other measurements the flow remained turbulent to much smaller Re_b similar to the result shown by (Jovanović *et al.*, 2011). The process in decreasing flow rate seems to be stochastic to some degree and it is very difficult to obtain a single representative curve. Such measurement for decreasing flow rate has been repeated several times, but a precisely reproducible curve was not found.

7.5.1 Discussion of transitional results in transitional channel flow

The present results clearly show an identical behavior of the smooth reference and the channel equipped with grooved plates. The structured plates behave like hydraulically smooth surfaces for the investigated Reynolds number range. Therefore, this result does not confirm previous findings concerning the drag reduction.

In the following differences in the experimental set-ups are pointed out and possible reasons for the differing results are concluded.

The geometry of the structures is differing between both studies. The grooves in the present study are slightly smaller and the curvature is V-shaped instead of rectangular. With these changes it was intended to increase the effectiveness of the structures. Despite these differences, it is remarkable, that there was no indication of a smallest difference in the wall shear stress between the reference and grooved surface in the present investigation.

The turbulence level for increasing flow rate showed no significant differences between the differing surface condition in the study by Jovanović *et al.* (2011) as well as in the present one. But the maximum peak of the turbulence level of naturally developing flow seems to occur at different Re_b in the two studies. In this particular case the difference is a result of the different calculation of the flow rate. In the measurements obtained in Erlangen the flow rate was calculated using the Bernoulli equation (cf. Spiegel (2013), Frohnapfel (2006)). This computation overestimates the flow rate compared to the calculation according to the DIN-norm by at least two to three percent. As shown in Appendix A.1 the use of inlet nozzles for low flow rates, where the flow is still laminar in the nozzle, overestimates the flow rate by another 5%. In sum a systematic *Re* deviation of approximately +7.5% results and this difference is indeed important. In the present study the nozzle has been calibrated with the orifice meter and can thus be used for low flow rates avoiding these systematic errors.

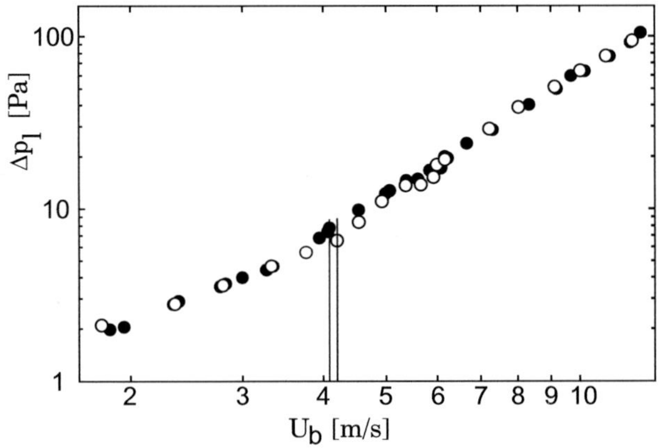

Figure 7.18: Comparison of the pressure drop in a complete smooth channel and a channel with a structured section in dependence of U_b (adapted from (Jovanović *et al.*, 2011)).

Additionally, the length of the tunnel plays an important role for the transition, since a longer running length enables a further growing of disturbances, which are inducing the transition. Taking this into consideration the transition occurs at similar values in the two facilities. If the corrected Reynolds number is calculated with the channel length as done by Jovanović *et al.* (2010b) the maximum turbulence occurs at $Re_x = 1.03 \times 10^6$ and almost identical value of $Re_x = 1.0 \times 10^6$ in present study.

Jovanović *et al.* (2011) explained the higher pressure loss in grooved upstream section of his channel as a "flow conditioning section". This characteristic is only visible in a single of measurement point in Jovanović *et al.* (2010a), which is depicted in Figure 7.8. In this particular case the velocity between the grooved and the reference case differed by approximately 2.5%. This is accentuated by the two vertical lines in the diagram 7.18, which is adopted from Jovanović *et al.* (2011). If we consider a quadratic relation of the pressure drop on U_b, an 2.5% higher velocity results in an increase of the pressure drop of 5% (assuming exactly the same air properties). Figure 7.8 compares the pressure drop along the test section for both cases. As expected for the flow of higher velocity (grooved surface) the pressure drop is larger on the upstream part of the channel, where the flow is laminar in both cases. On the downstream part of the channel the flow in the complete smooth channel develops turbulent at $x > 1.8m$, while the flow downstream the grooved surfaces remains laminar. Therefore, the measurement with higher velocity leads to less total pressure loss. This explanation to the different pressure drop on the upstream section seems much more plausible.

A general question concerning transition investigations in channel flow remains: Is a tunnel with an moderate aspect ration of 12 appropriate for this kind of testing?

The length of the channel may be large enough, that the transition-triggering disturbances are mainly self-generated by the side wall effects (secondary flow), as it was reported by Dean (1978). At a certain downstream distance the growing corner vortex structures observed in laminar flow through ducts interact with each other forming new disturbances. The very exact reproducibility of the onset of transition with the present set-up, may probably be a result of this. The observations of Arbeiter (2009) support such considerations (compare Chap. 2.2.3).

8 Spanwise oscillating walls implemented via dielectric elastomer actuators

This chapter presents the implementation of an active flow control technique in the present experimental set-up. The chosen technique of spanwise wall oscillation is promising in two aspects: a considerably large reduction of the mean streamwise skin-friction drag is obtained (Quadrio & Ricco, 2004) and the wall oscillations are a predetermined control strategy, meaning that the control action is defined a priori and does not depend on the flow state. Thus, compared to other active approaches, the wall oscillations are of moderate complexity, since neither complex control circuits nor sensors are required. The challenging task is the energy efficient implementation of large oscillating areas.

Recently, Gouder *et al.* (2013) presented an experimental study using a novel actuator concept for realization of spanwise wall movement: the dielectric electroactive polymers. The dielectric elastomer actuators exhibit several advantages in comparison to conventionally used approaches, where the walls are driven electrically or electromagnetically:

- They are capable of exerting in-plane, unidirectional oscillations.
- The lightweight devices comply with the requirements for flow control applications.
- The power consumption is low.

Although a preliminar study by Gouder *et al.* (2013) has shown that drag reduction by spanwise wall oscillations with dielectric polymer based actuators can be achieved, the technology was judged to be unattractive due to the fragility of the actuators. The lifetime of only a few minutes (Gouder, 2011) limited the possibility to experimentally study the DEA in the wind tunnel.

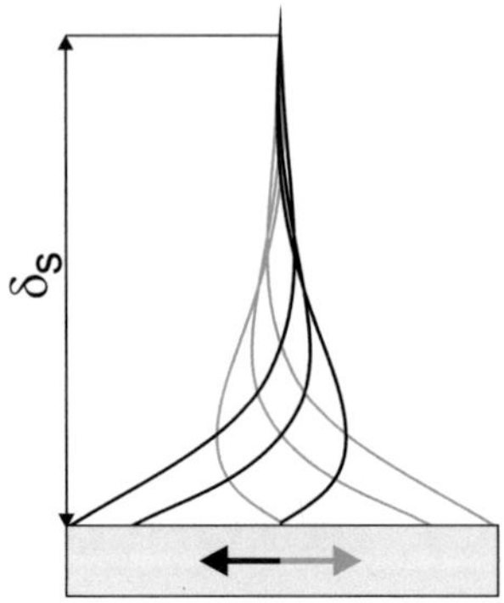

Figure 8.1: Stokes layer generated by cyclic movement of the wall.

The present study evolved from a cooperation with Davide Gatti, who advanced a new approach for the implementation of wall oscillations via dielectric electroactive polymers. Details of the material properties and the actuator design can be found in Gatti (2014b). The present wind-tunnel facility represents the perfect framework for testing these new devices for spanwise wall oscillations. While a detailed comparison of the measurement data to numerical simulations is presented in Gatti (2014b), the scope of the present work is to introduce and verify the experimental procedure.

In the following the concept of spanwise oscillating walls is introduced. Subsequently the working principle of a DEA and its application for the construction of a DEA actuator is shown. The challenges of implementing the DEA actuators in the experimental set-up are discussed. The measurement results are presented and the Chapter is completed with a discussion of the results.

8.1 Spanwise oscillating walls

An in-plane, oscillatory motion of the wall corresponds to the imposition of a periodic wall velocity u_w as follows:

$$u_w(t) = A_m \sin\left(\frac{2\pi}{T}t\right) \quad \text{with} \quad T = \frac{2\pi}{\omega} \tag{8.1}$$

In equation (8.1) A_m is the amplitude, T the period and ω the angular frequency of the oscillation. Consequently to the periodic spanwise wall velocity due to the no-slip condition, a periodic flow is created above the wall, which is confined to the so-called Stokes layer in a distance δ_S from the wall. The velocity profile for different phases of one period of the wall movement is sketched in Figure 8.1. The flow within the Stokes layer can be described by an analytic solution for the laminar case (Schlichting & Gersten, 2000):

$$u_w(x_2,t) = A_m \exp\left(-x_2\sqrt{\frac{\pi}{T\nu}}\right) \sin\left(\frac{2\pi}{T}t - x_2\sqrt{\frac{\pi}{T\nu}}\right) \qquad (8.2)$$

Drag reduction can be achieved, if spanwise wall oscillations are superposed to a turbulent wall-bounded flow. Vittori & Verzicco (1998) found that the main flow and the cross flow effectively interact only if the Reynolds number Re_δ, based on the thickness of the Stokes layer δ_S and the oscillation amplitude A_m, $Re_\delta = \frac{A_m \delta_S}{\nu}$ is smaller than 120. For this Reynolds number the cross flow in calm air would be laminar.

Since the seminal work of Jung *et al.* (1992), in which wall oscillations have been numerically proved to lead to turbulent drag reduction, a large number of numerical studies followed (Baron & Quadrio, 1996; Choi *et al.*, 2002; Duque-Daza *et al.*, 2012; Gatti & Quadrio, 2013; Quadrio & Ricco, 2003, 2004; Quadrio & Sibilla, 2000; Ricco & Quadrio, 2008; Touber & Leschziner, 2012; Yakeno *et al.*, 2014). Probably, the most comprehensive numerical study on the effect of oscillating walls was presented by Quadrio & Ricco (2004). DNS of turbulent channel flow at low *Re* have been exploited to investigate the optimal oscillations parameters and the important role of the oscillation period T was found. A optimal period in a range of T^+=100 to 120 was found, where T^+ is calculated with the friction velocity $u_{\tau,0}$ of the reference flow and the kinematic viscosity ν. A small number of experimental laboratory investigation confirmed the drag-reducing capabilities of wall oscillaitons (Choi & Clayton, 2001; Choi *et al.*, 1998; Di Cicca *et al.*, 2002; Gouder *et al.*, 2013; Ricco & Wu, 2004; Trujillo *et al.*, 1997).

Despite the large number of investigations the drag reduction mechanism is yet not understood. The different approaches to explain the mechanism are discussed by Gatti (2014b).

8.1.1 Dielectric electroactive polymers

Figure 8.2 illustrates the basic principle of a DEA. An isolating material, the dielectric, is sandwiched between two conductive compliant electrodes. This ensemble can be interpreted as a parallel plate capacitor. If an electric field is applied across the dielectric, the electrodes attract each other. The resulting forces exert a pressure on the dielectric, called the Maxwell pressure p_e:

$$p_e = \epsilon_0 \epsilon_r \frac{E^2}{h^2}. \tag{8.3}$$

The Maxwell pressure (Eq. (8.3)) depends on the the vacuum permittivity ϵ_0=8.85x10^{-12}F/m, the relative permittivity of the dielectric ϵ_r, the distance between the electrodes h and the potential difference between the electrodes E. In a standard plate capacitor, the stiff electrodes are fixed in a certain position and the deformation due to the Maxwell pressure is negligibly small. In a DEA the electrodes are of compliant material and applied on the elastic dielectric medium. In this context compliant means that the electrodes can significantly deform while remaining conductive. For the dielectric, a soft elastomeric material is usually used, which is highly deformable and can be regarded as incompressible. The Maxwell pressure deforms the isolating material. The elastomer is thinned and its surface area is expanded (Fig. 8.2).

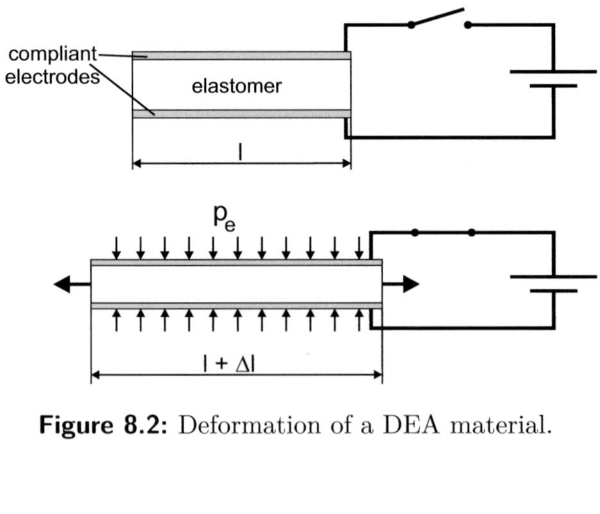

Figure 8.2: Deformation of a DEA material.

8.1.2 Working principle of dielectric elastomer actuators

To exploit the DEA for fluid mechanic applications, an in-plane oscillatory movement is required. In Figure 8.3 the main actuator arrangement is shown. A pre-stretched elastomeric membrane is clamped between two supporting frames. On opposite sides two pairs of compliant electrodes are applied onto the membrane. If the capacitor on the left side is electrically charged as shown in the upper illustration of Figure 8.3, the Maxwell pressure leads to an elongation of the underlying membrane and a movement towards the right side is generated. If this capacitor is discharged, while the one on the right side is loaded, a movement in the opposite direction is achieved. The pre-stretch of the elastomer is important to avoid an out of plane movement due to loss of tension of the membrane.

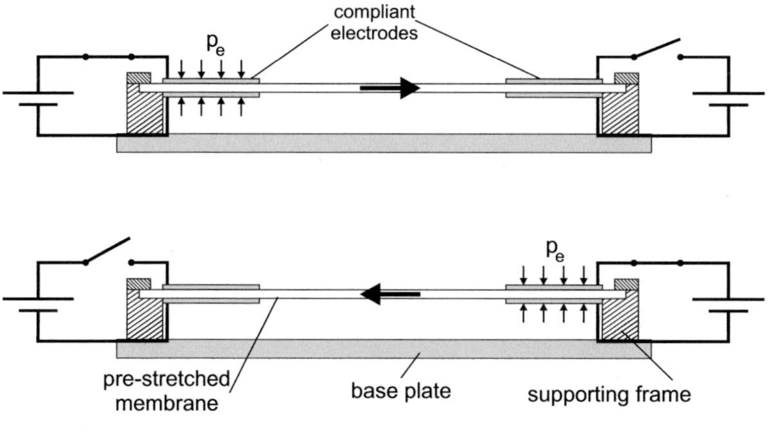

Figure 8.3: Working principle of a DEA.

The basic layout of the actuator for the wind tunnel testing is inspired by (Gouder *et al.*, 2013). The active membrane is not directly used to manipulate the flow. It is coupled with a second passive membrane which constitutes the surface wetted by the flow. A transfer element, which is glued on the membranes as depicted in Figure 8.4, directly couples the movement of the active and passive membrane. For a lightweight construction the transfer elements are made of balsa wood. The movement of the output shaft is illustrated in Figure 8.5. The passive surface is stiffened by a thin plastic plate to ensure a homogenous displacement of a large part of the actuated surface. An exploded view of the construction is depicted in Figure 8.6.

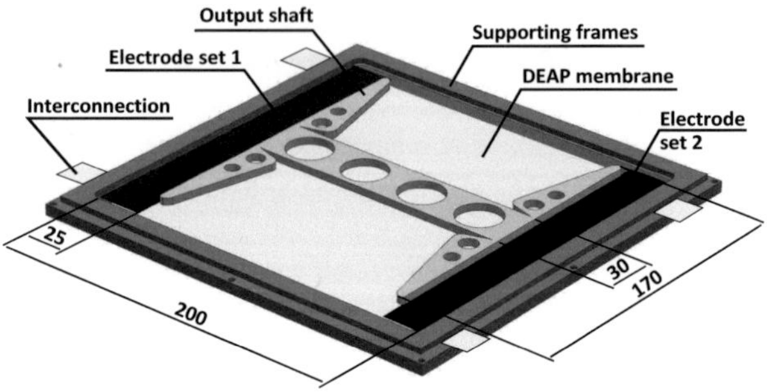

Figure 8.4: Active membrane of the DEA.

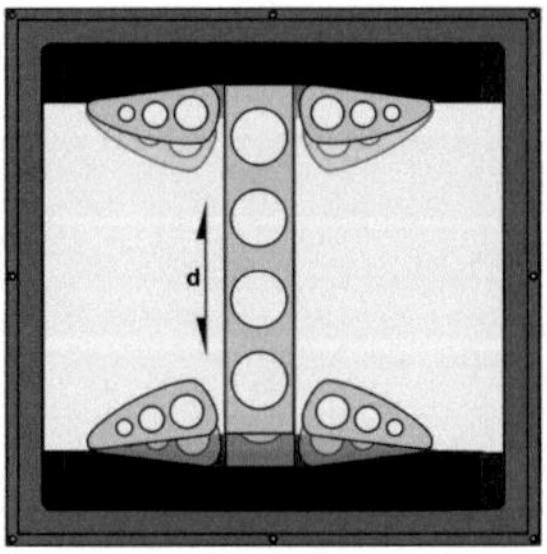

Figure 8.5: Illustration of the movement of the output shaft on the active surface of a DEA.

A large spanwise displacement is only achieved, when the actuator is operated at its in-plane mechanical resonance frequency. As a consequence, a given actuator design has a fixed working frequency, which depends on the stiffness of the membrane, the area of the membrane and the pre-stretch. The resonance frequency of actuators for the wind tunnel testing is additionally modified by the weight of the transfer elements and the stiffness of the surface membrane.

In the measurements two types of actuators were used. In Figure 8.6 the actuator type WT1 is shown, which has been developed by Gatti (2014b). It has an actuated length of 230mm in streamwise and spanwise direction.

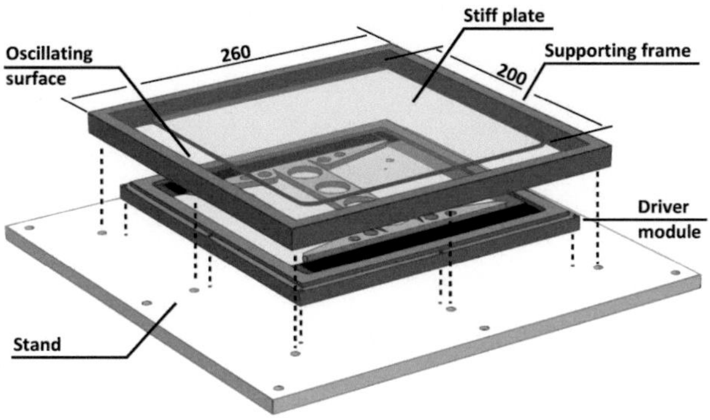

Figure 8.6: DEA of type WT1 (Gatti, 2014b).

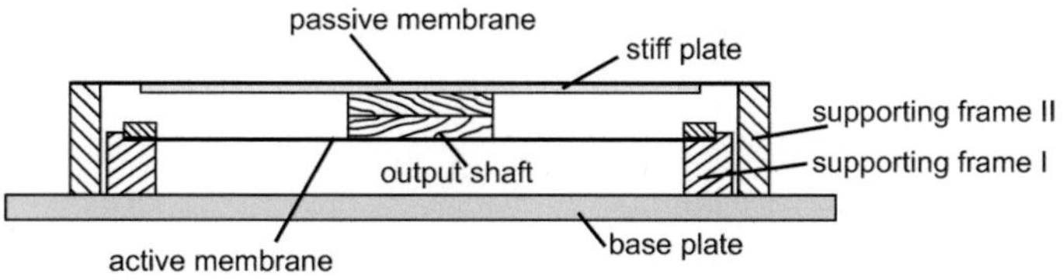

Figure 8.7: Sectional view in streamwise direction through a DEA.

The maximum planar displacement of about 4mm peak-to-peak is achieved at a frequency of 65Hz. The oscillating surface of the WT2 is enlarged to 270mm in spanwise and 270mm streamwise direction. The larger actuated surface enables larger displacement due to the reduced stiffness of the lager passive surface. The working frequency is decreased to 45Hz, the maximum displacement increased to 5mm. Gatti (2014b) exploits the empirical correlation proposed by Quadrio & Ricco (2004) to predict the Re range, in which the largest drag reduction is likely to occur. He predicts the range of Re of the best compromise between being sufficiently close to T^+_{opt} while still having a sufficiently large A^+_m to be $Re_b = 4000 - 8000$. A reduction of about 10% is predicted within this range.

8.1.3 Arrangement of experimental set-up

For the testing of the actuators a fully developed turbulent flow was required. In Chapter 5 this was obtained for a streamwise position of $x > 70 - 80H$.

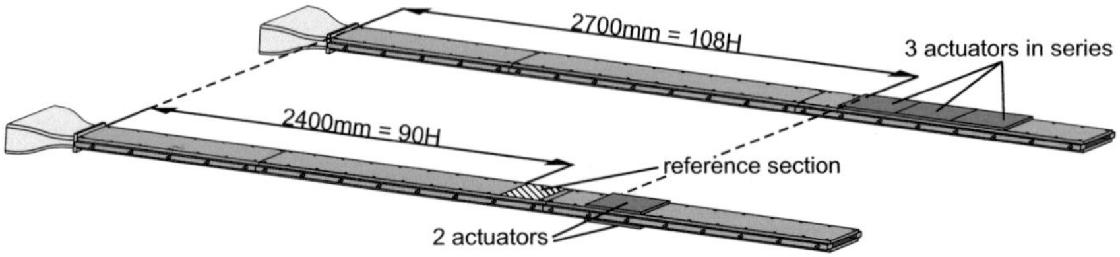

Figure 8.8: Arrangement of the test section for actuators in opposite wall and cascaded actuator configuration.

Therefore, the upstream channel sections (2450mm) were equipped with smooth aluminum plates and the 13% tripping was installed at the channel entrance. The leading edge of the actuators was positioned at the second pressure tap of the downstream section corresponding to $H > 108$.

The integration of such actuators in the wind tunnel test section was difficult. While the WT2 was especially designed for directly mounting to the sidewalls of the channel, the width of the WT1 was smaller than the channel width and an auxiliary plate was needed. The plate is made of plexiglas which ensures electric isolation and allows to look into the test section. The plexiglas plates were machined to ensure a planar intersection to the aluminum plates of the upstream channel section. Figure 8.9 shows the installation of the WT1 with the auxiliary frame on the bottom wall of the opened channel.

The height of the passive and active modules is adjusted by filling elements in order to ensure a smooth surface at the intersection between the actuator and the channel walls. These are thin layers of plastic film, interposed between the base plate and the supplying frames. Figures 8.10 (a) and (b) show the possible misalignments and the vertical adjustment of the frames support.

With the described arrangement enough space was left for an additional uncontrolled reference section, where the friction on a smooth surface was measured, between $x = 90H$ and 100H as shown in Figure 8.8. The smooth reference section is located upstream with a distance of 300mm to the beginning of the actuator to exclude an influence of the oscillations that might propagate downstream.

Two different actuator configurations were investigated as illustrated in Figure 8.8. In the first configuration two actuators of type WT1 were

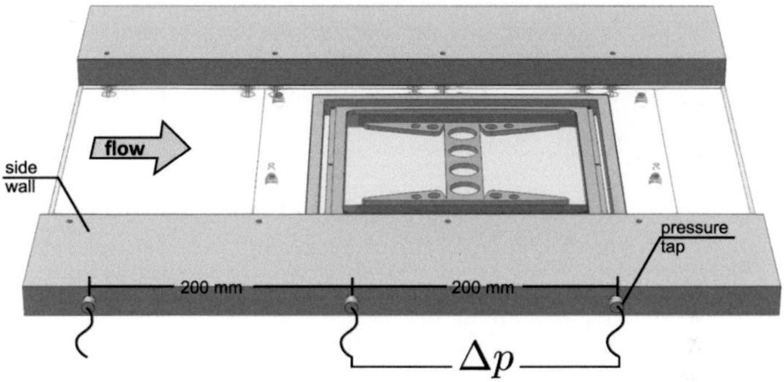

Figure 8.9: Integration of a WT1 on the bottom wall of the test section.

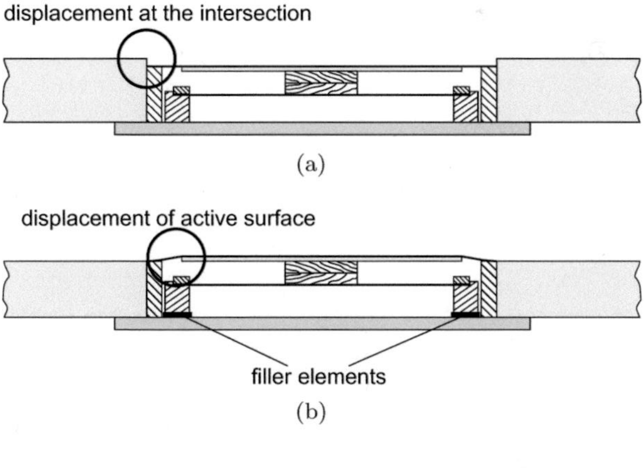

Figure 8.10: Vertical misalignment of the support frame of the passive membrane (a) and of the stiffening plate (b).

located on opposite walls. In the second configuration three actuators were cascaded only on the upper wall. A total streamwise actuator length of 900mm is generated. The actuated length was slightly smaller with 840mm, since the actuation is interrupted at the intersection between the devices.

8.1.4 Measurement strategy

The challenging aspect in testing the dielectric elastomer actuators was the short streamwise length of 230mm to 270mm of the actuated surface. Since the most effective operation of the actuators is predicted for $4000 < Re_b < 8000$, the pressure drop along such a small streamwise distance (200mm) is between 0.6Pa and 2Pa. To detect changes in the skin-friction with an accuracy better 1%, pressure changes of smaller 0.01Pa need to be measured.

From the reference measurement in turbulent flow of the riblet campaign, it can be concluded that $\mathrm{d}p/\mathrm{d}x$ is almost independent of the regarded channel length (compare Fig. 6.9). The deviation of the single data points along the channel length from a constant slope is extremely small, since the variation of the flow rate in time is very small for $Re_b > 5000$. It can be assumed that the measurement accuracy will be not be affected by the small regarded streamwise length, which is proofed by the measurement data shown at the end of this Chapter.

The pressure drop across both the actuator and the reference section have been measured repeatedly in six sessions of three minutes averaging time each, alternating actuator activation and deactivation. The change in τ_w is detected on the smooth and the oscillating surface. The smooth reference surface is needed to verify that the actuation does not influence the upstream flow. To allow a simultaneous measurement one Baratron sensor has been used for the reference section and the other one for the segment with the actuators.

In spite of the challenges of measuring small pressure drops across a small section, the testing of the active devices offers particular advantages that improve the measurement accuracy of $\tau_w/\tau_{w,0}$. Firstly, to investigate the effect of the wall oscillations, only the actuator needs to be switched on and off. A reassembly of the test section for the reference case is not required. The switch between actuator on and off leads to almost the same favorable situation for the measurement accuracy as the direct comparison in Chapter 6. The measurement can be conducted within short time and thereby deviations of the ambient quantities are negligibly small. Also time dependent deviations of the pressure sensors, both for flow rate measurement and pressure drop measurement become very small, since a whole test series can be conducted within one day. Additionally, at this time the wind tunnel instrumentation was improved to the the final step of development with a more precise determination of calculation of the air density and viscosity.

The calculation of the measurement uncertainty has been adapted for the mentioned boundary conditions and the uncertainty intervals shown in the following section represent this result.

8.2 Measurement results

8.2.1 Measurement of reference case

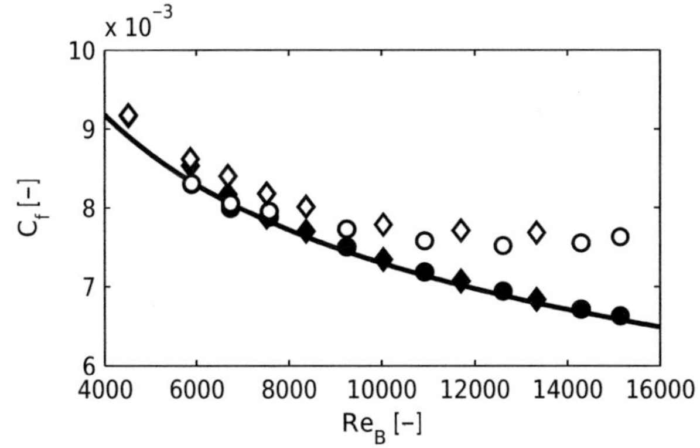

Figure 8.11: Determination of c_f on the smooth reference section (filled symbols) and the section with the deactivated actuator (open symbols) from simultaneous measurement. Circles depict results of the fist measurement series. Diamonds depict results of the second measurement series.

The data in Figure 8.11 show a comparison of the skin-friction coefficient over the smooth reference surface to the Dean's law, which was found to describe very accurately the relation among c_f and Re for the present channel in Chapter 5.2. Additionally, the diagram includes data for the actuator section without oscillation with two WT1 actuators in opposite wall configuration. Two different measurement series are included in the plot. For the smooth channel section the skin friction coefficient agrees very well with both Dean's law and with the measurements presented in Chapter 5.2. The membrane surfaces show good agreement for small Re only. With increasing Re the skin friction on the actuator section increases compared with the smooth reference. The actuator behaves as a rough surface. This can be caused by:

- improper adjustment of the actuator in the test section,
- improper adjustment of the active surface,
- stimulation of vibrations of the membrane, leading to an out of plane movement (without actuation).

The deviation from the smooth reference is found for both test series. However, there is a significant difference between the two measurement series on the actuator section. This difference originates from the fact that the passive membrane of one actuator broke after the first test series. Although the actuator seemed to be identical after having been repaired and it was carefully reassembled into the channel section, the skin friction measurements over the actuator surface indicates that the intersection was not as good as before. This shows how sensitive the results are against the smallest misalignment in the actuator section.

8.2.2 DEA in opposite wall configuration

In the first test series the actuator of type WT1 has been tested in the opposite wall configuration (compare Fig. 8.8). The measurement results in Figure 8.12 show the relative change in the wall friction against Re_b with the actuator switched on and off

$$\Delta\tau_w/\tau_{w,off} = (\tau_{w,off} - \tau_{w,on})/\tau_{w,off}. \tag{8.4}$$

Each measurement point was repeated three times as can be observed in Figure 8.12. The figure includes the results of two different test series with the same actuator type. Additionally, mean values of the three single measurement points for each Re are included. The second test series showed a decrease in the skin friction reduction, which in the previously shown

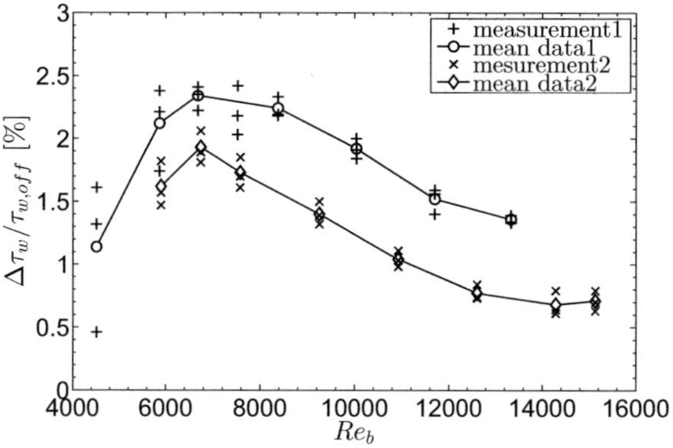

Figure 8.12: Measurements of $\Delta\tau_w/\tau_{w,off}$ in dependence of Re_b for two different test series (+,x). The open symbols show the mean value of three reproductions of the measurement.

reference measurement was identified as a result of improper adjustment of the sensors in the test section. Therefore, the second test series is not taken into account for the following discussion. Note that the first test series may also be influenced by a similar effects but of lower magnitude. Scatter in the data is very low considering the extremely small changes in the skin friction. With increasing Re the scatter reduces from $\pm 0.4\%$ to $\pm 0.2\%$.

The skin friction reduction increases to a maximum of slightly more than 2% at $Re_b = 7000$. For higher Reynolds numbers the skin friction reduction decreases. In Gatti (2014b) the agreement to numerical results is discussed.

Figure 8.13 verifies the measurement uncertainty of the data for $\Delta\tau_w/\tau_{w,off}$. The shaded areas show the 90%-confidence interval, which has been computed by extending the standard deviation of the mean of the pressure signals during the measurement Gatti (2014b) (light grey) and the result of uncertainty estimation in Chapter 4 (black), which is comparable to 99.7% confidence interval. For moderate and large Re the 90% confidence interval is about one third smaller than the 99.7% confidence value, which indicated good agreement between the different approaches to calculate the uncertainty. For $Re_b < 6000$ the scatter in the data is larger

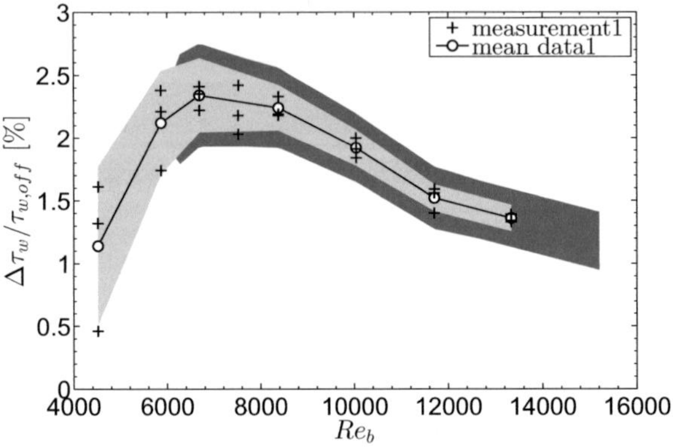

Figure 8.13: Comparison of the uncertainty in the determination of $\Delta\tau_w/\tau_{w,off}$: The measurement data are compared to a 90%-confidence interval (light grey) computed in Gatti (2014b) and the result of uncertainty estimation in Chapter 4 (black).

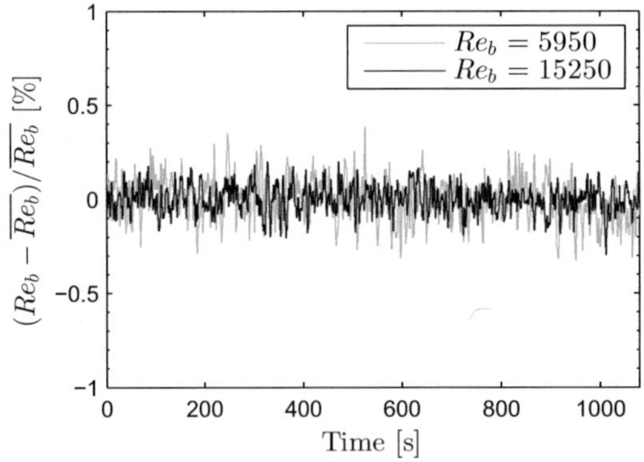

Figure 8.14: Percentage fluctuation of Re_b around the mean value. For the lower and upper end of the measurement range. The actuator is activated and deactivated in time intervals of 180s.

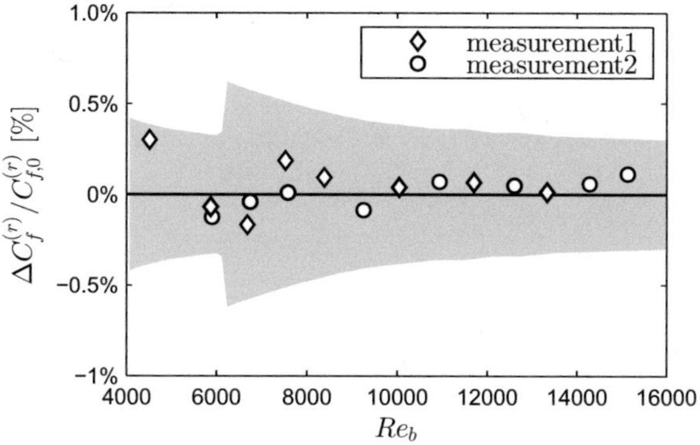

Figure 8.15: Influence of the activation of the DEA on the skin friction on the upstream reference section. Data for both test series are included. The grey shaded area depicts the uncertainty interval according to the analysis in Chapter 4.

than predicted by the present uncertainty calculation. This calculation predicts an increase of the accuracy due to the switch to the more accurate pressure sensor at the flow meter (compare Chapter 4.2.6). In contrast for low Re the scatter of the data is not dominated by the accuracy of the flow meter in this particular measurement. Two other sources lead to an increase of the scatter instead. Firstly, the measurement point at $Re_b = 4600$ is slightly below $Re_b = 4800$, at which a fully turbulent flow can be ensured throughout both downstream sections of the channel. This increases the scatter in the flow rate, as was found in Chapter 5 and is depicted in detail in Figure 6.12. Secondly, fluctuations of the flow rate particularly decrease the measurement accuracy in the DEA testing, since the determination of the pressure drop is based only on a small streamwise length and additionally only one measurement point. The latter aspect was discussed in Section 8.1.4. It can be concluded that the measurement accuracy is appropriate for $Re_b > 6000$ to identify changes in $\Delta\tau_w/\tau_{w,off}$ in the order of 0.4%. The measurement accuracy increases with the Re_b and allows to detect changes of 0.2% for $Re_b > 9000$.

Figure 8.14 shows the measurements of the flow rate for the complete measurement time of 18min, in order to prove that the reduced pressure loss along the actuator is not a result of a change in flow rate. The actuator is activated and deactivated in time intervals of 180s. Exemplary measurements for $Re_b = 5950$ and $Re_b = 15250$ are depicted, which represent the measurement range in which a high confidence in the result previously was confirmed from the uncertainty quantification. The variation in the flow rate is very small, for the mean value between the three minutes intervals it is far less than 0.1%.

Complementary, Figure 8.15 shows the change in the skin friction on the smooth reference section $\Delta c_f^{(r)}/c_{f,0}^{(r)}$, while the actuator was activated and deactivated on the simultaneously measured section. On the reference section the flow is confirmed to be unaffected by the actuation. Scatter in the data is less than 0.3%, which supports the previously shown result of the uncertainty consideration.

8.2.3 DEA in cascaded configuration

For the measurements shown in this paragraph two changes in the set-up were made. The configuration of the actuators has been changed to the cascaded configuration, Figure 8.8. The main advantage is a longer actuated length in streamwise direction. The second change was the usage of the new WT2 actuator to obtain larger displacement. Unfortunately, the large deformations of the active membrane resulted in more frequent failures, which reduced the durability of the WT2 actuators.

Since the actuators were only arranged on the upper channel wall, the skin friction reduction is assumed to double for a virtual two-sided configuration in order to enable comparisons with the first test series:

$$\Delta\tau_w/\tau_{w,off} \cong 2(\tau_{w,off} - \tau_{w,on})/\tau_{w,off}. \tag{8.5}$$

Gatti (2014b) showed the validity of this approach in a numerical simulation with the same configuration.

Figure 8.16 shows the results of $\Delta\tau_w/\tau_{w,off}$ for the WT2, which is corresponding to the total actuated length. Unfortunately the test series could not be completed since the membranes of the actuators broke after four measurements. It can be construed that a even higher $\Delta\tau_w/\tau_{w,off}$ could have been achieved at higher Re. Although only the incomplete data set can be compared with the WT1 configuration, the measurement indicate a considerable influence of the transient length on the effectiveness of the technique.

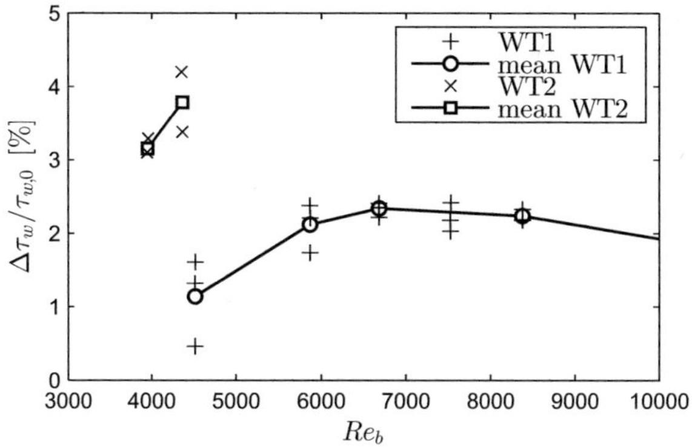

Figure 8.16: Reduction of skin friction for the test of cascaded WT2 actuators.

8.2.4 Influence of limited spanwise actuator length on skin friction reduction

Analogously to the analysis carried out in Chapter 5.4.3, the drag reduction is corrected by taking into account the deviation of the flat duct from a ideal two-dimensional flow. Additionally, the fact that only a part of the channel width has been actively controlled is also taken into account, assuming that no drag reduction occurs in the non-actuated surface area. Furthermore, the actuation is not homogenous over the complete actuated surface. While the stiffened surface area of the passive membrane (200x200mm for both actuator types) oscillates homogeneously, the displacement linearly decreases to zero towards the supporting frames. Therefore, the fraction of the actuated length without stiffening is considered with a mean value half of the maximum displacement. For the WT1 a correction factor for the change in the skin friction of k_{WT1}=1.4 resulted and for the WT2 type of k_{WT2}=1.28. The corrected values for the maximum $\Delta\tau_w/\tau_{w,off}$ are 3.2% for WT1. The WT2 case with cascaded actuators is more complex, since the supporting frames between adjacent actuators interrupt the actuated surface and cannot be corrected analogously.

8.3 Discussion of the DEA results

The experiments of the present study were able to proof the applicability of the dielectric elastomer actuators for flow control. In reproducible measurements a reduction of the skin friction on the actuated surface was shown. There are two main reasons for this achievement. On the one hand, the actuators built by Gatti (2014b) were durable and allowed to run a whole experimental campaing with manageable effort; on the other hand the present set-up provided the ideal framework for the actuator testing. In fact, although the reduction was comparably small with maximum three percent, the accuracy of the instrumentation allowed to precisely investigate the effect in dependence of Re. It was shown that for $Re_b > 6000$ the changes in $\Delta\tau_w/\tau_{w,off}$ of the order of 0.4% were statistically significant. The measurement accuracy even increases with the Re_b and allowed to detect changes of 0.2% for $Re_b > 9000$.

The careful measurement of the reference case highlighted an increase of the skin friction on the deactivated actuator in comparison to the reference case for $Re_b > 7000$ (see Fig. 8.11). For $Re_b > 9000$ the reduction of τ_w by the wall oscillation does not compensate the increase of the skin friction due to the presence of the actuator. However, at $Re_b \approx 6500$, where the maximum of $\Delta\tau_w/\tau_{w,off}$ was achieved, there is a holistic drag reduction, meaning the skin friction is below the value on the smooth reference section.

The DEA developed by Gatti (2014b) can have comparably long life-time. In particular, the WT1 types were still working after more than ten hours of measurement. The WT1 never showed a defect of the active membrane, but a passive membrane broke twice. The WT2 type turned out to be more fragile. The achievement of large displacement reduced the lifetime of the active membrane, which impeded a comprehensive study. However, the few measurement results obtained with the cascaded configuration indicate that the effectiveness of the spanwise oscillations technique can significantly be enhanced by increasing the actuated streamwise length.

9 Summary and conclusions

The present work deals with high-precision skin-friction measurements of various active and passive surfaces in a newly built wind-tunnel facility and is comprised of three consecutive parts. The first part contains an in-depth report of both design and construction of the test-rig. The measurement accuracy and the precisely adjustable test conditions have been verified in the second part by means of a hydraulically smooth reference surface. Subsequently, various passive and active surfaces were tested for their drag-reduction capabilities as compared to this reference. Finally, a discussion of the achieved insights, drawn conclusions and remarks on possible future efforts with this new test rig closes the present work.

The first part of the work has been devoted to the description of the experimental facility and its construction, comprising the wind tunnel, the channel flow test section and the instrumentation. A channel flow test section with an aspect ratio of $AS = 12$ was chosen to approximate the favorable test conditions of a two-dimensional flow. The skin friction is determined indirectly through measures of the pressure drop in streamwise direction. A wind tunnel of blower type has been chosen, as it allows to built-up a comparatively high pressure and is compatible to be equipped with precise flow rate instruments. The modular construction, which allows the facility to reliably cover a large operation range and to be highly extendable, was realized according to the engineer standards DIN EN ISO 5167-2 (2003) and DIN 24163 (1985). Air was chosen as the working medium in order to allow the application of active flow controllers, some of which cannot operate in liquids. However, this choice implies the demanding task to measure extremely small variations of small quantities as compared to measurements in liquids.

Despite this adverse choice of air, it was the main goal of the tunnel and instrumentation design process to achieve at least comparable accuracy as reported for oil and water tunnels. In particular, the test section has been equipped with pressure transducers of highest accuracy available on the market, which made a detection of extremely small pressure variations (thus wall shear stresses) possible. In addition, a precise determination of the flow rate has been obtained by the use of inlet nozzles and orifice meters. For the latter, parameter range and sensitivity studies have proven the reproducibility of the measurements and revealed an extended working range of nozzle flow meters below their critical flow rate. The detailed estimation of measurement uncertainty and error propagation of preliminary tests identified an insufficient reproducibility ($\pm 1\%$) of the flow rate as the main remaining weakness for the determination of skin friction changes.

To reduce this remaining amount of uncertainty, the resolution of skin friction changes was further improved by three different approaches during experimentation. First, the impact of ambient noise on the measurement data was identified and these effects on the measurement results were excluded. Second, environmental influences like state-variable changes were minimized by an additional consideration of the varying air humidity. Compensation for this effect further reduced the uncertainty by 0.2%. Third, the experimental procedure has been modified for the respective test cases in so as to ensure highest achievable precision for the comparison of the flow controlled surfaces and the reference surfaces. These modifications have been discussed extensively and their impact on the measurement accuracy was reflected in the experimental results. In the riblet campaign skin friction changes of 0.4% could be resolved at moderate Reynolds numbers of 10000 by a comparison of reference and test surface in two simultaneously measured sections of the channel. Note that larger Reynolds numbers revealed even higher accuracy. In the DEA study the measurement accuracy of 0.4% was achieved at significantly smaller Reynolds numbers of 6000. For $Re_b > 9000$ it even increased to 0.2%. This improvement was possible, since the actuators were switched on and off repeatedly. Consequently, the relative difference between either case was less affected by any drift of boundary conditions.

Prior to any flow control experiments, an extensive study of the smooth reference surface was performed, where excellent agreement with literature was demonstrated for both laminar and turbulent flow. Furthermore, an accurate examination of the flow development was possible, which provided a refined specification of the development length of a turbulent channel flow. While in the literature (Durst *et al.*, 1998) for a streamwise distance larger than 60 channel heights the development was reported to be completed, it turned out from the present study that a streamwise distance of 70 to 80H should be exceeded. However, this longer development length implies further skin friction changes of 0.3%, which corresponds to the resolution limit of with the present set-up. The resulting overall uncertainty of approximately 0.4% proved that wind tunnels can indeed operate as accurate as today's most accurate oil tunnels (Bechert *et al.*, 1992). Furthermore, to improve the comparability between experimental studies and numerical simulations, the spanwise distribution of the wall shear stress has been investigated. This data serves as the basis for a correction of the three dimensional effects generated by the channel sidewalls, which will be conclusively discussed in the next section.

The applicability of the above-introduced test rig for highly accurate skin friction (reduction) measurements was then demonstrated by means of two passive and one active flow control surface. The main emphasis was placed on the study of riblet structures and micro grooves. In addition, novel dielectric elastomer actuators (DEA) have been applied in the present tunnel. The development of the DEA has initially been an independent research project (Gatti, 2014b). Since these active flow control actuators perfectly fit into the present wind tunnel facility, it was extremely beneficial for both projects to study the achievable flow control success of DEA in the present experimental set-up. Insights and drawn conclusions from all three drag-reduction approaches will be summarized below, as well.

9.1 Comparability of experimental and numerical data

Already the literature research at the beginning of the work showed the lack of an analytic formulation for the friction coefficient c_f in a turbulent channel flow in dependence of Re. For the experiments where the channel is approximated by a flat duct, the aspect ratio of the width to height of the duct AS is an additional parameter influencing c_f. The experimental approaches, that are found in literature (Dean, 1978; Johnston, 1973; Zanoun *et al.*, 2009) correlate measurement data for different ranges of Re_b. For the calculation of c_f a uniform spanwise velocity distribution is assumed, which does not consider the presence of the the side walls. The considerable deviation from a two-dimensional flow was confirmed by the comparison with numerical results (Hoyas & Jiménez, 2008; Iwamoto *et al.*, 2002; Moser *et al.*, 1999). The empirical formulas from all experimental approaches were based on the same functional relation to correlate the measurement data $c_f = A \cdot Re_b^{-0.25}$. In dependence of the considered Re range the literature sources found different results for the constant A, which shows that any of the correlations is not appropriate for a large Re range. The present data confirmed the approach by Dean (1978) to describe very accurately the dependence of the skin friction coefficient on Re for a channel of moderate aspect ratio ($AS = 12$) at $Re_b = 4000-24000$. These boundary conditions are representative for the most channel flow experiments for drag reduction investigations of flow control. Although this Re range contains the low Reynolds numbers at which DNS are conducted, the comparability of the experimental studies to the numerical results is decreased due to the different spanwise velocity distribution and the additional friction generated by the side walls, which affect the obtained skin friction changes of the flow control applications. To improve this situation measurements were conducted to quantify the deviation from a uniform spanwise wall shear distribution for the present channel. In the considered measurement range, the distribution of the wall shear stress was obtained to be nearly independent of Re. As a result an empirical correction function for skin friction change $\delta\tau_w/\delta\tau_{w,0}$ for an $AS = 12$ channel was derived. Assuming the main change of the velocity distribution in dependence of the aspect ratio of flat ducts was the extension of the nearly uniform spanwise distribution, which was found around the channel center, the correction function was generalized for channels of $AS > 5$. A much better comparability of channel flow experiments of different AS and experimental and numerical data is hypothesized.

9.2 Riblets

The riblet study was the benchmark-test for the experimental set-up, since the riblet structures have been extensively studied (Bechert *et al.*, 1997; Grüneberger & Hage, 2011; Viswanath, 2002; Walsh, 1982; Walsh & Lindemann, 1984; Walsh & Sellers, 1988). In particular the experimental research by Bechert *et al.* (1997) with a special oil tunnel facility designed for the detection of skin friction changes is from unmatched quality. It was demonstrated that the present set-up very well reproduces these results and that scatter in the data of both facilities is comparable, although both facilities utilize completely different frameworks for the testing. Bechert's oil tunnel (Bechert *et al.*, 1992) of a rectangular cross section directly measures τ_w with a shear stress balance, while the present set-up indirectly determines the wall shear stress from the streamwise pressure drop of an airflow. It is the first time that riblets have been measured with such accuracy in an air flow.

On that basis it was enabled to exploit the benefit of an air flow facility to study the flow development on riblet surfaces over long streamwise distances. After a minimal onset length a constant reduction of the skin friction on the riblet surface was obtained: The estimate by Walsh (1990) that the amount of drag reduction is independent of the streamwise position is confirmed.

The study identified the leading edge of the riblets plate to be crucial. Although the investigated structures were very small, the increase of the pressure drop at the leading edge was significant. This observation is in contradiction to Bechert *et al.* (1997) who found an abrupt begin of the riblet structures not to affect the result of $\tau_w/\tau_{w,0}$. Controversial findings in the literature (Bechert *et al.*, 1997; Squire & Savill, 1986, 1987; Walsh, 1990) are indicated to be more related to the design of the experiment than to the riblets. In wind tunnel facilities the riblet height is negligible small compared to the dimensions of the facility, while in a channel flow experiment, the protruding riblets are relatively larger compared to characteristic height. Thus, in channel flow experiments, the measurement is much more sensitive to the small blockage by the protruding fins. Additionally, shear stress balances, which were usually used for riblet studies, do not provide a streamwise resolution of the wall shear stress distribution and the comparably small leading edge effect may disappear within the integral value of the test plate on the balance.

9.3 Microgrooves

Even though disillusioning and contradictory at first glance, the careful analysis of the micro groove data reveals clear and conclusive insights. It has been demonstrated in the present work that the studies of presumingly turbulent flow by Frohnapfel *et al.* (2007a); Jovanović *et al.* (2010b) in fact were neither turbulent nor fully developed. The different wall shear stress on upstream and downstream section of the channel at low *Re* was found to be attributed to the laminar turbulent development of the flow. Hence, the present analysis clearly depicts the strong dependence of the significance of measurement results on the measurement uncertainty and the detailed analysis of the reference case.

For the study in naturally developing flow a modified groove geometry was chosen. Instead of rectangular, V-shaped grooves were used with the intention to increase the transition delay found by Jovanović *et al.* (2011). However, the surfaces with streamwise aligned grooves were not able to delay the transition in the present measurement, but the skin friction was equal to that of hydraulically smooth plates. Finally, the reason for the discrepancies to the measurement of Jovanović *et al.* (2011) cannot definitely be clarified. Nevertheless, the present study is superior in regard to the measurement accuracy as well as the number and reproducibility of the measurements.

Instead the fundamental question arises whether the transition of naturally developing flow in a channel flow facility of this type in generally can be delayed, if the manipulation is not focusing on the probably dominant factor to trigger transition: the secondary motions contaminating the wall effects. The present study agrees to this statement by Dean (1978) and is confirmed by the observations of Arbeiter (2009). Furthermore, the similar Reynolds number at which the highest turbulence level was obtained in the present study and the investigation by Jovanović *et al.* (2011) points in this direction. A pipe flow experiment would be beneficial, since a circular cross section avoids the side wall effects. However, the transition mechanisms in inner flows and a zero pressure gradient boundary layer are not the same. Therefore, contrary to turbulent flows, the results from a pipe flow cannot directly be transferred to a boundary layer flow.

The micro grooves idea needs to return to the origin. The misinterpretation of experimental data led to the conclusion (Frohnapfel *et al.*, 2007a), that the optimum scaling of the grooves width is in the order of a viscous length scale. The theoretical idea as well as the numerical simulations (Frohnapfel, 2006) were focusing on dimensions of the viscous sublayer ($x_2^+ = 5$). The ineffectiveness of the grooves probably is not originating from the absence of one-component turbulence within the groove, but the near wall flow is not effectively influenced by the grooves. The rectangular groove geometry must lead to failure, since the interaction of the flow with the sharp angle flanks of the grooves generates a high local friction. Eventual benefit inside the groove is compensated, as was discussed by Walsh (1990) in the context of optimization of riblet geometries. The actual groove scaling is so small, that the roughness represented by the structures is negligible for a large Reynolds number range. In comparison with the criterion of Nikuradse (1933) the size of micro grooves is diminutive. Also the recently identified effects on turbulent flow by structures of even smaller scales by Grüneberger & Hage (2011) indicate the micro grooves to be without significant effect on a turbulent flow. Due to the alignment in flow direction and the surface embedded shape the grooves are of less effect than the small fins orientated in cross flow direction in Grüneberger's investigation ($h^+ = 0.5$). In general, the groove geometry is in contradiction to the riblets design. The experimental investigations by Walsh (1982) showed that neither V-grooves nor convex semicircular structures, which are very similar to the optimized groove geometry suggested in (Frohnapfel, 2006), achieved a reduction of the skin friction drag at a scaling in order of the viscous sublayer and larger.

The idea of the recently investigated similar structures, but of considerably larger scaling Daschiel (2014) is based on the finding by Eckert & Irvine (1955). Eckert found the co-existence of laminar and turbulent flow in pipes of non-circular cross sections, which were suggested for a transition delay in Eckert & Irvine (1960). This approach is adapted in Daschiel (2014) where the scaling of the grooves is not of the order of typical surface morphology, but represents a modification of the global shape of the cross section. From the view point of the present author, the development of theses large-scale structures for transition delay is the only remaining starting-point to achieve a drag reduction on base of the idea of streamwise aligned groove structures.

9.4 Dielectric elastomer actuators

In the DEA study the implementation of an active flow control technique into the channel test section has been successfully demonstrated. These novel actuator type to generate an oscillating wall is quite fragile and conductivity of the working fluid is problematic. Therefore, to study actuators of such type is only possible in an air flow experiment. The short streamwise length of the actuated surface represented a very challenging aspect in testing the DEA. The most effective operation of the actuators is found in the Re range of $4000 < Re_b < 8000$, in which the pressure drop along such a small streamwise distance (200mm) was between 0.6Pa and 2Pa. The changes due to the actuation were even two orders less.

Despite these challenging aspects the measurement strategy, which enabled to compare the deactivated and activated devices alternating, turned out beneficial for the significance of the measurement with the active device. The experimental set-up verified the calculated uncertainty to allow the detection of changes in the skin-friction with an accuracy of 0.4% at $Re_b = 6000$. The significance even increased with Re_b and allowed to detect changes of 0.2% for $Re_b > 9000$, which demonstrates the maximum accuracy to be almost independent of the considered streamwise distance.

With the dielectric elastomer actuators a reduction of the friction about $\Delta\tau_{w,max} = 3\%$ has been precisely quantified. It was also shown that the skin friction was reduced below the value of a simultaneously measured hydraulically smooth surface, on which the skin friction very well agreed to the correlation by Dean (1978).

The actuators developed by Gatti (2014b) represent substantial progress. In contrast to the dielectric elastomer based actuators shown by Gouder *et al.* (2013), who reported the life-time of the devices to be few minutes, Gatti's WT1 actuators sustained ten hours of testing without damage. Through the insufficient durability of the WT2 actuators this test series was interrupted and remained incomplete. However, in preliminary testing to check the proper working of the set-up with actuators, significantly larger drag reduction has been observed. It is not adventurous to predict that this technique can have twice the effectiveness than what could be recorded in the present study.

10 Outlook

The first part of the outlook aims to expose the variability of the present set-up to inspire future research projects using the facility. The second part suggests practical examples to continue research related to the present study.

10.1 Prospects of the facility

The following section discusses further improvement of the facility and highlights the potentials of the large operative range and variability of the facility in different configurations.

With regard to the absolute accuracy of the skin friction determination a calibration of the orifice meters would be desirable. While the detection of changes in $\tau_w/\tau_{w,0}$ is of very high accuracy, the uncertainty of the skin friction detection could be further improved. In Chapter 4 it was elaborated, that the largest contribution to the uncertainty of the determination of the skin friction originates from the flow rate meter. The systematic uncertainty of the flow coefficient of the orifice meter was identified to be about half of the total uncertainty of the flow rate measurement. Note that the agreement to Dean's law implies that there is an uncertainty in the flow coefficient, but the effectively resulting error is very small. An improvement of the accuracy of flow meters can be achieved by a calibration using an instrument, for instance an inlet nozzle, of high specified accuracy. The required test rig for a direct comparison of different flow rate meters through simultaneous measurement has already been tested during the present work (see Appendix A.1).

If future projects continue to use the wind tunnel configuration with the channel test section of the present work, it might be useful to install a smaller blower with maximum flow rate about $2500\text{m}^3/\text{h}$ and the maximum pressure built-up at lower flow rates. This would be advantageous to achieve even more constant flow rates at very low Re. Additionally the minimum flow rate could be reduced to $Re_b \approx 1500$, which would enlarge the operation range in laminar flow. The flow rate instrumentation would only require a smaller orifice disk to be adapted to the new operation range.

In order to attend to the recent numerical results by Daschiel (2014), it would be straight-forward to combine the wind tunnel with a pipe flow test section. The enormous length (18m) of the laboratory in Karlsruhe represents perfect boundary conditions to use the facility in combination with a long pipe test section. This pipe test section would also take advantage of all previously mentioned suggestions for improvement of the experimental set-up.

One potential option is to build a test section of comparably large cross section in the range of 0.5m× 0.3m ($W \times H$) to generate a wind tunnel design like a "pressure driven Eiffel tunnel". Since this type of tunnel is lacking at the institute in Karlsruhe. It might be even useful to equip the facility with a larger ventilator for this purpose. The Eiffel tunnel configuration would allow to conduct experiments for classical aerodynamic research.

10.2 Studies of channel flow

In order to improve the comparability of numerical and experimental studies of channel flow an empirical approach based on measurements of the spanwise distribution of the velocity and the wall shear stress for turbulent flow was made. However, in the experimental investigations it is not possible to study the ideal two-dimensional reference case and thus we cannot compare experiment results of the approach by the flat duct with the uniform velocity distribution. In the numerical studies this would be possible. The two-dimensional channel flow with periodic boundary conditions has been intensively studied (Hoyas & Jiménez, 2008; Iwamoto *et al.*, 2002; Moser *et al.*, 1999). Additionally, numerical studies allow to investigate the flow in a flat duct and to study the influence of secondary flows, which was not possible in the present experimental approach. Recently, numerical studies by Vinuesa (2013) and Vinuesa *et al.* (2014) showed results for flat ducts including such data. However, the Re was quite low and the aspect ratio was limited to $AS = 7$, which is still too low to approximate a two-dimensional flow. It would be worth to invest the time for DNS at moderate Reynolds numbers of 5000 to 6000 of duct flows with $AS \geq 10$ for a direct comparison with the two-dimensional results. This would allow to verify and refine the experimental approach of the present study. The outcome would be meaningful for any channel flow experiment both numerical and experimental.

The essential question which came along with the studies of naturally developing channel flow is the releasing mechanism of the transition to turbulence in a flat duct. The literature could not answer the question to which degree the sidewall effects contaminate the cross section. Neither a specification of the streamwise length, nor the Re range, where these effects begin to dominate the transition process is given. An essential question is whether an higher aspect ratio would limit the contamination of the flow by the sidewall effects to a distinct area and the channel center would remain unaffected. This is indicated in measurements by Nishioka & Masahito (1984), where a channel with a large aspect ratio of 27.4 was used. Although the tunnel is quite long, $L/H = 400$, laminar flow is retained up to $Re_b = 10650$. The contamination of the flow by vortices originating from the sidewalls is not mentioned there. However, only the 60% around the center of spanwise velocity distribution is shown by Nishioka *et al.* (1975) and the sidewall region is omitted. It is speculated that this spanwise area is mainly influenced by the sidewall effects. Anyhow, this large aspect ratio channel exhibits a region in the channel center, which is not affected by the secondary flows. It can be deduced that the aspect ratio may play a more important role for transition process than initially considered.

To study the development of the corner vortices, the available channel could be equipped with plexiglas plates on top. Additionally, a seeding generator at the entrance of the nozzle for a flow visualization would be required. This would allow to study the occurrence and development of the contamination of the spanwise flow field in dependence of Re and the running length. Eventually, the present channel would be more convenient for transition testing in a shortened configuration, where the sidewall effects have less streamwise distance to develop.

10.3 Channel flow studies at high Re

The experimental set-up allows to operate the channel test section at significantly higher Re. With the available diffuser at the outlet of the channel $Re_b > 40000$ has been achieved. If the suggestion to install a blower with larger pressure built-up is followed, the Reynolds number can be increased to more than twice this value. In the following two ideas, which are based on this high Re configuration, are described to contribute to quite fundamental questions.

Recently Gatti (2014b) discussed the decrease in drag reduction effectiveness of flow control techniques with increasing Reynolds number, which has been obtained in numerical as well as experimental studies. The experiment described in the following could contribute to verify the relation of the drag reduction on Re, which was proposed by Gatti (2014b).

From the present riblet study it can be observed that an accuracy of determining skin frictions changes of 0.4% would be ensured for $Re_b \geq 10000$. If two similar riblet structures at different scaling could be produced, with an optimum spacing about $Re_b = 10000$ and $Re_b = 80000$ the effect of the Reynolds number on the drag reduction could be studied very precisely. Note that it would not be required to produce high performance riblets, but to achieve the same relative fabrication quality of both riblet surfaces.

Additionally, the high Re set-up would allow to investigate the skin friction coefficient for a wide Reynolds number range with an accuracy, which has not been achieved before. It would enable to study the change of the skin friction coefficient at higher Reynolds number. Thereby one important issue needs to be considered: the change in the spanwise velocity profile with increasing Re, which was indicated in Chapter 5.4. The function previously used (Dean, 1978; Johnston, 1973; Zanoun *et al.*, 2009) $c_f = A \cdot Re^{-0.25}$ was found to be unsuitable to describe the dependence of c_f on wide range of Re. Instead even A is supposed to depend on Re.

The gain would be an accurate empirical correlation for $c_f = f(Re_b)$ for a channel with an aspect ratio of 12. In optimal case the outcome could be combined with result of the previously suggested study to quantify the difference of numerical and experimental channel flows to achieve a correlation for $c_f = f(Re_b, AS)$.

Bibliography

Arbeiter, F. 2009. *Untersuchungen von Kühlgasströmungen bei transitionellen Reynoldszahlen im Einlaufbereich von Minikanälen.* Ph.D. thesis, Universität Karlsruhe.

Arbeiter, F., Gordeev, S., Heinzel, V., Ihli, T., Leichtle, D., Möslang, A., & Slobotchouka, V. 2008. Features and optimization approaches of the entrance section cooling gas flow of IFMIF High Flux Module. *Fusion Engineering and Design*, **83**, 1536–1542.

Bailey, S. C. C., Hultmark, M., Monty, J. P., Alfredsson, P. H., Chong, M. S., Duncan, R. D., Fransson, J. H. M., Hutchins, N., Marusic, I., McKeon, B. J., Nagib, H. M., Örlü, R., Segalini, A., Smits, A. J., & Vinuesa, R. 5. 2013. Obtaining accurate mean velocity measurements in high Reynolds number turbulent boundary layers using Pitot tubes. *J. Fluid Mech.*, **715**, 642–670.

Baron, A., & Quadrio, M. 1996. Turbulent drag reduction by spanwise wall oscillations. *Appl. Sci. Res.*, **55**, 311–326.

Bechert, D. W. 1995. Calibration of Preston tube. *AIAA Journal*, **34**(1), 205–206.

Bechert, D. W, Hoppe, G., van der Hoeven, J. G. T., & Makris, R. 1992. The Berlin oil channel for drag reduction research. *Exp. Fluids*, **12**, 251–260.

Bechert, D.W. 1988. Some concepts and facilities for drag reduction research. *Third Turbulent Drag Reduction Working Party, ONERA-Chatilion, Paris.*

Bechert, D.W., & Bartenwerfer, M. 1989. The viscous flow on surfaces with longitudinal ribs. *J. Fluid Mech.*, **206**, 105–209.

Bechert, D.W., Bruse, M., Hage, W., van der Hoeven, J.G.T., & Hoppe, G. 1997. Experiments on drag-reducing surfaces and their optimization with an adjustable geometry. *J. Fluid Mech.*, **338**, 59–87.

Blasius, H. 1913. Das Ähnlichkeitsgesetz bei Reibungsvorgängen in Flüssigkeiten. *Forsch.-Arb. Ing.-Wes., Heft 131.*

Bradshaw, P. 2000. A note on "critical roughness height" and "transitional roughness". *Phys. Fluids*, **12**(6), 1611–1614.

Bradshaw, P., & Pankhurst, R. C. 1964. The design of low speed wind tunnels. *Progr. Aeronaut. Sciences*, **5**, 1–69.

Bradshaw, Peter, Gregory, Nigel, & Britain, Great. 1961. *The determination of local turbulent skin friction from observations in the viscous sub-layer.* HM Stationery Office.

Byrne, J., Hatton, A.P., & Marriot, P.G. 1970. Turbulent flow and heat transfer in the entrance region of a parallel wall passage. *Proc. Instn. Mech. Engrs.*, **184**(39), 697–710.

Carlson, D. R., Widnall, S. E., & Peeters, M. F. 1982. A flow visualization study of transition in plane Poiseuille flow. *J. Fluid Mech.*, **121**, 487–505.

Cattafesta, Louis N., & Sheplak, Mark. 2011. Actuators for active flow control. *Ann. Rev. Fluid Mech.*, **43**, 247–272.

Choi, H., Moin, P., & Kim, J. 1991. On the effect of riblets in fully developed laminar channel flows. *Phys. Fluids*, **A3**(1892).

Choi, H., Moin, P., & Kim, J. 1993. Direct numerical simulation of turbulent flow over riblets. *J. Fluid Mech.*, **255**, 503–539.

Choi, J.-I., Xu, C.-X., & Sung, H. J. 2002. Drag reduction by spanwise wall oscillation in wall-bounded turbulent flows. *AIAA J.*, **40**(5), 842–850.

Choi, K.-S., & Clayton, B.R. 2001. The mechanism of turbulent drag reduction with wall oscillation. *Int. J. Heat a. Fluid Flow*, **22**, 1–9.

Choi, K-S., DeBisschop, J.R., & Clayton, B.R. 1998. Turbulent boundary-layer control by means of spanwise-wall oscillation. *AIAA J.*, **36**(7), 1157–1162.

Colebrook, C.F. 1939. Turbulent flows in pipes with particular reference to the transition between the smooth- and rough-pipe laws. *J. Inst. Civil Eng.*, **11**, 133–156.

Coleman, H. W., & Steele, W. G. 2009. *Experimentation, validation, and uncertainty analysis for engineers.* John Wiley & Sons, Inc.

Czichos, H., & Hennecke, M. 2004. *Hütte, Das Ingenieurswesen 32. Auflage*. Springer.

Daschiel, G. 2014. *Strategies to reduce friction losses and their implications for the energy efficient design of internal flow domains*. Ph.D. thesis, Karlsruher Institut für Technologie.

Davies, S. J., & White, C. M. 1928. An experimental study of the flow of water pipes of rectangular section. *Proc. R. Soc. London A*, **119**, 92–107.

Dean, B, & Bhushan, B. 2010. Shark-skin surfaces for fluid-drag reduction in turbulent flow: a review. *Phil. Trans. R. Soc. A*, **368**, 4775–4806.

Dean, R. B. 1978. Reynolds number dependence of skin friction and other bulk flow variables in two-dimensional rectangular duct flow. *J. Fluids Engin.*, **100**, 215–223.

Di Cicca, G.M., Iuso, G., Spazzini, P.G., & Onorato, M. 2002. PIV investigation of turbulent boundary layer manipulated by spanwise wall oscillations. *J. Fluid Mech.*, **467**, 41–56.

DIN 24163. 1985. *Ventilatoren*. German Standard.

DIN EN ISO 5167-1. 2003. *Durchflussmessung von Fluiden mit Drosselgeräten in voll durchströmten Leitungen mit Kreisquerschnitt - Allgemeine Grundlagen und Anforderungen*. International Standard (German version).

DIN EN ISO 5167-2. 2003. *Durchflussmessung von Fluiden mit Drosselgeräten in voll durchströmten Leitungen mit Kreisquerschnitt - Blenden*. International Standard (German version).

Duchmann, A., Grundmann, S., & Tropea, C. 2013. Delay of natural transition with dielectric barrier discharges. *Exp. Fluids*, **54**, 1461.

Duque-Daza, C.A., Baig, M.F., Lockerby, D.A., Chernyshenko, S.I., & Davies, C. 2012. Modelling turbulent skin-friction control using linearised Navier-Stokes equations. *J. Fluid Mech.*, **702**, 403–414.

Durst, F., Fischer, M., Jovanović, J., & Kikura, H. 1998. Method to set up and investigate low Reynolds number, fully developed turbulent plane channel flows. *Trans. ASME I: J. Fluids Eng.*, **120**(3), 496–503.

Durst, F., Ray, S., Ünsal, B., & Bayoumi, O. A. 2005. The development length of laminar pipe and channel flows. *Trans. ASME*, **127**(6), 1154–1160.

Eckert, E. R. G., & Irvine, T. F. 1955. Simultaneous turbulent and laminar flow in ducts with noncircular cross section. *J. Aeronaut. Sciences*, **22**, 65.

Eckert, E. R. G., & Irvine, T. F. 1960. Laminar skin friction and heat transfer on flat plates with wedge-shaped grooves in flow direction. *Trans. ASME*, **82**(4), 325–332.

Eggels, J.G.M., Unger, F., Weiss, M.H., Westerweel, J., Adrian, R.J., Fiedrich, R., & Nieuwstadt, F.T.M. 1994. Fully developed turbulent pipe flow: a comparison between direct numerical simulation and experiment. *J. Fluid Mech.*, **268**, 175–209.

Ferriss, DH. 1965. *Preston tube measurements in turbulent boundary layers and fully developed pipe flow.* HM Stationery Office.

Fink, V. 2014. *Numerical investigation of the impact of riblets on turbulent channel flow.* Master's thesis, Karlruher Institut für Technologie.

Frohnapfel, B. 2006. *Flow control of near-wall turbulence.* Ph.D. thesis, Universität Erlangen-Nürnberg.

Frohnapfel, B., Jovanović, J., & Delgado, A. 2007a. Experimental investigations of turbulent drag reduction by surface-embedded grooves. *J. Fluid Mech.*, **590**, 107–116.

Frohnapfel, B., Lammers, P., Jovanović, J., & Durst, F. 2007b. Interpretation of the mechanism associated with turbulent drag reduction in terms of anisotropy invariants. *J. Fluid Mech,*, **577**, 457–466.

Frohnapfel, B., Lammers, P., & Jovanović, J. 2007c. *New results in numerical and experimental fluid mechanics.* Springer. Chap. The role of turbulent disspation for flow control of near-wall turbulence, pages 268–275.

Frohnapfel, B, Lammers, P., Jovanović, J., & Delgado, A. 2008. Hydraulic model of the skin friction reduction with surface grooves. *IUTAM Symposium on Flow Control and MEMS*, **7**, 427–431.

Fulachier, L., Djenidi, L., & Anselmet, F. 1987. Boundary layers on longitudinally grooves "Riblets". *Association Aeronautique at Astronautique de France*, **PB 88**(205620).

Gad-el Hak, M. 2000. *Flow control – passive, active and reactive Flow Management.* Cambridge University Press.

Gallagher, J. A., & Thomas, A. S. W. 1984. Turbulent boundary layer characteristics over streamwise grooves. *AIAA Paper 84-2185.*

Garcia-Mayoral, R., & Jiménez, J. 2011. Hydrodynamic stability and the breakdown of the viscous regime over riblets. *J. Fluid. Mech.*, **678**, 317–347.

Gatti, D. 2014a. *Mean velocity profile of turbulent channel flows at* $Re_B =$ 13650. Tech. rept. Technische Universität Darmstadt.

Gatti, D. 2014b. *Turbulent skin-friction drag reduction via spanwise wall oscillations.* Ph.D. thesis, Technischen Universität Darmstadt.

Gatti, D., & Quadrio, M. 2013. Performance losses of drag-reducing spanwise forcing at moderate values of the Reynolds number. *Phys. Fluids*, **25**(125109).

Gatti, D., Haus, H., Matysek, M., Frohnapfel, B., Tropea, C., & Schlaak, H. F. 2014. The dielectric breakdown limit of silicone dielectric elastomer actuators. *Appl. Phys. Lett.*, **104**(5), 052905.

Gouder, K. 2011. *Turbulent friction drag reduction using electroactive polymer surfaces.* Ph.D. thesis, Imperical College, London.

Gouder, K., Potter, M., & Morrison, J.F. 2013. Turbulent friction drag reduction using electroactive polymer and electromagnetically driven surfaces. *Exp. Fluids*, **54**(1), 1441.

Grüneberger, R., & Hage, W. 2011. Drag characteristics of longitudinal and transverse riblets at low dimensionless spacings. *Exp. Fluids*, **50**(2), 363–373.

Hartnett, J. P., Koh, J. C. Y., & McComas, S. T. 1962. A comparison of predicted and measured friction factors for turbulent flow through rectangular ducts. *ASME J. Heat Trans.*, **84**, 82–88.

Head, M. R., & Rechenberg, I. 1962. The Preston tube as a means of measuring skin friction. *J. Fluid Mech.*, **14**, 1–17.

Herwig, H., & Voigt, M. 1995. Turbulent entrance flow in a channel: An asymptotic approach. *Lecture Notes in Physics*, **442**(51-58).

Herwig, Heinz. 2008. *Strömungsmechanik.* Vieweg+Teubner Verlag.

Hoagland, L. C. 1960. *Fully developed turbulent channel flow in straight rectangular ducts - secondray flow, its cause and effct on the primary flow.* Ph.D. thesis, Massachusetts Institute of Technology.

Holderried, S. 2011. *Inbetriebnahme des neuen Windkanals "Emmy".* Bachelor's thesis, Technische Universität Darmstadt.

Hoyas, S., & Jiménez, J. 2008. Reynolds number effects on the Reynolds-stress budgets in turbulent channels. *Phys. Fluids*, **20**(10), 101511.

Huebscher, R. G. 1947. Friction in round, square and rectangular ducts. *Heating, Piping and Air Conditioning*, **19**, 455–467.

Iwamoto, K., Suzuki, Y., & Kasagi, N. 2002. Reynolds number effect on wall turbulence: toward effective feedback control. *Int. J. Heat Fluid Flow*, **23**(5), 678–689.

Jiménez, J., Hoyas, S., Simens, M. P., & Mizuno, Y. 2010. Turbulent boundary layers and channels at moderate Reynolds numbers. *J. Fluid Mech.*

Johnston, J. P. 1973. The suppression of shear layer turbulence in rotating systems. *ASME J. Fluids Engin.*, **95**(229-236).

Jovanović, J. 2004. *The statistical dynamics of turbulence.* Springer.

Jovanović, J., Frohnapfel, B, Lienhart, H., & Delgado, A. 2010a. Experimental investigation of viscous drag reduction by flow control of laminar to turbulent transition using microgroove surface pattern. *ETMM8, Marseille, France*, 773–780.

Jovanović, J., Frohnapfel, B., & Delgado, A. 2010b. Viscous drag reduction with surface-embedded grooves. *Turbulence and interactions, notes on numerical fluid mechanics and multidisciplinary design*, **110**, 191–197.

Jovanović, J., Frohnapfel, B., Srikantharajah, R., Jovanovitc, D., Lienhart, H., & Delgado, A. 2011. Microflow-based control of near-wall fluctuations for large viscous drag reduction. *Microfluid Nanofluid*, **11**, 773–780.

Jung, W.J., Mangiavacchi, N., & Akhavan, R. 1992. Suppression of turbulence in wall-bounded flows by high-frequency spanwise oscillations. *Phys. Fluids A*, **4 (8)**, 1605–1607.

Kennedy, J. F., Hsu, S.-T., & Liu, J.-T. 1973. Turbulent flows past boundaries with small streamwise fins. *J. Fluid Mech.*, **99**, 605–616.

Kim, J., Moin, P., & Moser, R. 1987. Turbulence statistics in fully developed channel flow at low Reynolds number. *J. Fluid Mech.*, **177**, 133–166.

King, R. A., & Breuer, K. S. 2001. Acoustic receptivity and evolution of two-dimensional and oblique disturbances in a Blasius boundary layer. *J. Fluid Mech.*, **432**, 69–90.

King, R. A., & Breuer, K. S. 2002. Oblique transition in a laminar Blasius boundary layer. *J. Fluid Mech.*, **453**, 177–200.

Kline, S. J., Reynolds, W. C., Schraub, F. A., & Runstadler, P. W. 1967. The structure of turbulent boundary layers. *J. Fluid Mech.*, **30**, 741–773.

Kurz, A., Goldin, N., King, R., Tropea, C., & Grundmann, S. 2013. Hybrid transition control approach for plasma actuators. *Exp. Fluids*, **54**(11), 1–4.

Lammers, P., Jovanović, J., & Durst, F. 2006. Numerical experiments on wall turbulence at low Reynolds number. *J. Therm. Science*, **10**(2), 33–62.

Laufer, J. 1951. Investigation of turbulent flow in a two-dimensional channel. *NACA Rept. 1053.*

Lee, S.-J., & Choi, Y.-S. 2008. Decrement of spanwise vortices by a drag-reducing riblet surface. *J. Turb.*, **9**(23), 1–15.

Lee, S.-J., & Lee, S.-H. 2001. Flow field analysis of a turbulent boundary layer over a riblet surface. *Exp. Fluids*, **30**, 153–166.

Liandrat, J. 1988. Laminar boundary layer over riblets. A first step towards the turbulent case. *Third Turbulent Drag Reduction Working Party, ONERA-Chatilion, Paris.*

Luchini, P. 1993. Viscous sublayer analysis of riblets and wire arrays. *Appl. Sci. Res.*, **50**, 255–266.

Luchini, P., & Quadrio, M. 2006. A low-cost parallel implementation of direct numerical simulation of wall turbulence. *J. Comp. Phys.*, **211**(2), 551–571.

Luchini, P., Manzo, F., & Pozzi, A. 1991. Resistance of a grooved surface to parallel flow and cross-flow. *J. Fluid Mech.*, **228**, 87–109.

Lumley, J. 1977. Drag reduction in two-phase and polymer flows. *Phys. Fluids*, **20**(10), 64.

Lumley, J., & Kobu, I. 1985. Turbulent drag reduction by polymer additives: a survey. *In:* B. Gampert – Springer Verlag (ed), *The influence of polymer additives on velocity and temperature fields (IUTAM symposium Essen 1984)*.

Marusic, I., McKeon, B. J., Monkewitz, P. A., Nagib, H. M., Smits, A. J., & R., Sreenivasan K. 2010. Wall-bounded turbulent flows at high Reynolds numbers: Recent advances and key issues. *Phys. Fluids*, **22**.

McKeon, B. J., Li, J., Jiang, W., Morrison, J. F., & Smits, A. J. 1999. Further observations on the mean velocity distribution in fully developed pipe flow. *J. Fluid Mech.*, **501**(135).

McKeon, B. J., Li, J., Jiang, W., Morrison, J. F., & Smits, A. J. 2003. Pitot probe corrections in fully developed turbulent pipe flow. *Meas. Sci. Tech.*, **14**, 1449–1458.

McLean, J., George-Falvy, D., & Sullivan, P. 1987. Flight test of turbulent skin-friction reduction by riblets. *Pages 408–424 of: Proc. RAeS Int. Conf. on Turbulent Drag Reduction by Passive Means, London*, vol. RAeS 2. Royal Aeronautical Society.

Mehta, R.D., & Bradshaw, P. 1979. Design rules for small low-speed wind tunnels. *Aero. J.*, **73**, 443.

Moin, P., & Mahesh, K. 1998. DIRECT NUMERICAL SIMULATION: A tool in turbulence research. *Ann. Rev. Fluid Mech.*, **30**, 539–578.

Monty, J. P. 2005. *Developments in smooth wall turbulent duct flows*. Ph.D. thesis, The University of Melbourne.

Moody, L. F., & Princeton, N. J. 1944. Friction factors for pipe flow. *Trans. ASME*, **66**(8), 671–684.

Moreau, E. 2007. Airflow control by non-thermal plasma actuators. *J. Phys. D: Appl. Phys.*, **40**, 605–636.

Moser, R. D., Kim, J., & Mansour, N. N. 1999. Direct numerical simulation of turbulent channel flow up to Re= 590. *Phys. Fluids*, **11**(4), 943–945.

Nagib, H. M., Vinuesa, R., Noorani, A., Lozano-Duràn, a., El Khoury, G., Schlatter, P., & Fischer, P. F. 2013. Direct numerical simulations of variable-aspect-ratio turbulent duct flows at low to moderate Reynolds numbers. *presented at the TSFP8 in Poitiers, France.*

Nikuradse, J. 1932. Gesetzmäßigkeiten der turbulenten Strömung in glatten Rohren. *Forschungsheft 356, Beilage zu Forschung auf dem Gebiete des Ingenieurwesens, Ausgabe B Band 3, VDI-Verlag Berlin.*

Nikuradse, J. 1933. Strömungsgesteze in Rauhen Rohren. *VDI Forsch. 361.* (Engl. Trans. 1950. Laws of flow in rough pipes. NACA TM 1292).

Nishioka, M., Iida, S., & Ichikawa, Y. 1975. An experimental investigation of the stability of a plane Poiseuille flow. *J. Fluid Mech.*, **72**, 731–751.

Nishioka, M., M., & Masahito, A. 1984. Some observations of the subcritical transition in plane Poiseuille flow. *J. Fluid Mech.*, **150**, 441–450.

Nitsche, W., & Brunn, A. 2006. *Strömungsmesstechnik.* Springer.

Orszag, S. A., & Patera, A. T. 1983. Secondary instability of wall-bounded shear flows. *J. Fluid Mech,*, **38**, 347–385.

Patel, V. C. 1965. Calibration of the Preston tube and limitations on its use in Pressure Gradients. *J. Fluid Mech.*, **23**, 185–208.

Pope, A., & Harper, J.J. 1966. *Low-speed wind tunnel testing.* John Wiley & Sons, Inc.

Pope, S.B. 2000. *Turbulent Flows.* Cambridge University Press, Cambridge.

Preston, J.H. 1954. The determination of turbulent skin friction by means of Pitot tubes. *J. R. Aeronaut Soc.*, **14**, 109–121.

Quadrio, M., & Ricco, P. 2003. Initial response of a turbulent channel flow to spanwise oscillation of the walls. *J. Turb.*, **4**(7).

Quadrio, M., & Ricco, P. 2004. Critical assessment of turbulent drag reduction through spanwise wall oscillation. *J. Fluid Mech.*, **521**, 251–271.

Quadrio, M., & Sibilla, S. 2000. Numerical simulation of turbulent flow in a pipe oscillating around its axis. *J. Fluid Mech.*, **424**, 217–241.

Reader-Harris, M. J., & Sattary, J. a. 1996. The orifice plate discharge coefficient equation - the equation for ISO 5167-1. *In Proc. of 14th North Sea Flow Measurement Workshop, Peebles, Scotland, East Kilbride, Glasgow, National Engineering Laboratory.*

Reif, W.-E. 1978. Protective and hydrodynamic function of the dermal skeleton of elamolranshs. *Neues Jahrbuch für Geologie und Paleontologie*, **157**, 133–141.

Reif, W.-E., & Dinkelacker, A. 1982. Hydrodynamics of the squamation in fast swimming sharks. *Neues Jahrbuch für Geologie und Paleontologie*, **164**, 184–187.

Reynolds, O. 1883. An experimental investigation of the circumstances which determine whether the motion of water shall be direct or sinuous, and of the law of resistance in parallel channels. *Phil. Trans. R. Soc. A*, **174**, 935–982.

Ricco, P., & Quadrio, M. 2008. Wall-oscillation conditions for drag reduction in turbulent channel flow. *Int. J. Heat Fluid Flow*, **29**, 601–612.

Ricco, P., & Wu, S. 2004. On the effects of lateral wall oscillations on a turbulent boundary layer. *Exp. Therm. Fluid Sci.*, **29**(1), 41–52.

Sareen, A., Deters, R. W., Henry, S. P., & Selig, M. S. 2011. Drag reduction using riblet film applied to airfoils for wind turbines. *AIAA Paper 2011-588*, **588**.

Savill, A. M. 1986. Effects of turbulent boundary bayer structure of longitudinal riblets alone and in combination with outer layer devides. *Fourth International Symposium on Flow Visualization, Paris.*

Sawyer, W. G., & Winter, K. G. 1987. An investigation of the effect on turbulent skin friction of surfaces with streamwise grooves. *Turbulent Drag Reduction by Passive means Conference, London.*

Schickel, T. 2012. *Beeinflussung einer Kanalströmung durch mikrostrukturierte Oberflächen.* Bachelor's thesis, Technische Universität Darmstadt.

Schlichting, H. 1934. Laminare Kanaleinlaufströmung. *ZAMM*, **13**, 260–263.

Schlichting, H. 1959. *Grenzschichttheorie.* G. Braun, Karlsruhe.

Schlichting, H., & Gersten, K. 2000. *Boundary-layer theory.* Springer, Berlin.

Schultz, M. P., & Flack, K. A. 2013. Reynolds-number scaling of turbulent channel flow. *Phys. Fluids*, **25**.

Skinner, G. T:. 1951. *Mean speed measurements in two-dimensional, incompressible, fully-developed turbulent channel flow.* M.Phil. thesis, California Institut of Technology.

Spiegel, E. A. 2013. *Laminare Strömungskontrolle durch in die Oberfläche eingebettete Rillen.* Master's thesis, Universität Erlangen-Nürnberg.

Spudat, W. 1981. *Erstellung einer Dokumentation über den Mehrzweckversuchsstand MZV des Lehrstuhls für Strömungslehre.* Studienarbeit, Universität Erlangen-Nürnberg.

Spurk, H., & Aksel, N. 2006. *Strömungslehre.* Springer.

Squire, L. C., & Savill, A. M. 1986. High-speed riblets. *European Meeting on Turbulent Drag Reduction, Lausanne.*

Squire, L. C., & Savill, A. M. 1987. Some experiences of riblets at transonic speeds. *Turbulent Drag Reduction by Passive Means Conference, London.*

Stokes, G. 1851. On the effect of the internal friction of fluids on the motion of pendulums. *Trans. Cambridge Phil. Soc.*, **9**, 8–106.

Sutardi, & Ching, C. Y. 2001. Effect of tube diameter on Preston tube calibration curves for the measurement wall shear stess. *Exp. Therm. Fluid Science*, **24**, 93–97.

Toms, B. A. 1948. Some observations on the flow of linear polymer solutions through straight tubes at large Reynolds numbers. *Proc. Int. Congr. on Rheology, North-Holland, Amsterdam*, **2**, 135–141.

Touber, E., & Leschziner, M.A. 2012. Near-wall streak modification by spanwise oscillatory wall motion and drag-reduction mechanisms. *J. Fluid Mech.*, **693**, 150–200.

Tropea, C., Damaschke, N., & Nobach, H. 2003. *Grundlagen der Messtechnik I.* Shaker Verlag.

Tropea, C., Yarin, A., & Foss, J. F. (eds). 2007. *Handbook of Experimental Fluid Mechanics.* Springer.

Trujillo, S.M., Bogard, D.G., & Ball, K.S. 1997. Turbulent boundary layer drag reduction using an oscillating wall. *AIAA Paper 97-1870.*

VDI. 2002. *VDI-Wärmeatlas.* Springer-Verlag.

Vinuesa, R. 2013. *Synergetic computational and experimental studies of wall-bounded turbulent flows and their two-dimensionality.* Ph.D. thesis, Illinois Institute of Technology.

Vinuesa, R., Noorani, A. N., Lozano-Durán, A., El Khoury, G.E.K., Schlatter, P.and Fischer, P F, & Nagib, H.M. 2014. Aspect ratio effects in turbulent duct flows studied through direct numerical simulation. *J. Turb.*, **15**, 677–706.

Viswanath, P.R. 2002. Aircraft viscous drag reduction using riblets. *Prog. Aerospace Sci.*, **38**, 571 – 600.

Vittori, G., & Verzicco, R. 1998. Direct simulation of transition in an oscillatory boundary layer. *J. Fluid Mech.*, **371**, 207–232.

von Kármán, T. 1930. mechanische Ähnlichkeit und Turbulenz. *In Proc. Third Int. Congr. Applied Mechanics, Stockholm*, 85–105.

Walsh, M. J. 1982. Turbulent boundary layer drag reduction using riblets. *AIAA Paper 82-0169.*

Walsh, M. J. 1983. Riblets as a viscous drag reduction technique. *AIAA Journal*, **21**(4), 485–486.

Walsh, M. J. 1990. *Viscous drag reduction in boundary layers.* Progress in Astronautics and Aeronautics, vol. 123. AIAA. Chap. Riblets, pages 203–261.

Walsh, M. J., & Lindemann, A. M. 1984. Optimization and application of riblets for turbulent drag reduction. *AIAA Paper 84-0347.*

Walsh, M. J., & Sellers, W. L. 1988. Riblet drag reduction at flight conditions. *AIAA Paper 88-2554.*

Walsh, M. J., & Weinstein, L. M. 1978. Drag and heat transfer on surfaces with small longitudinal fins. *AIAA Journal*, **17**(7), 770–771.

White, F. M. 1974. *Viscous fluid flow.* McGraw Hill.

Yakeno, A, Hasegawa, Y, & Kasagi, N. 2014. Modification of quasistreamwise vortical structure in a drag-reduced turbulent channel flow with spanwise wall oscillation. *Phys. Fluids*, **26**(8).

Zanoun, E-S., Nagib, H., & Durst, F. 2009. Refined c_f relation for turbulent channels and consequences for high-Re experiments. *Fluid Dynamic Research*, **41**(2), 021405.

List of Figures

List of Tables

A Appendix

A.1 Extension of the operation range to small flow rates for inlet nozzles

This appendix discusses the extension of the measurement range of inlet nozzles and describes the calibration conducted in the present work.

In the second part of DIN 24163 (1985) the lower limit for the use of inlet nozzles is specified with $Re_D \geq 100000$. An extension of the measurement range is desirable especially for small nozzles. While for nozzles of large diameter the pressure difference becomes minimal and in consequence the measurement accuracy decreases, small nozzles still have large enough pressure differences. In a simple test two inlet nozzles were installed serially in the wind tunnel: one at the inlet and one at the outlet of the facility. The latter sucked the air from the settling chamber. The two nozzles would definitely have to measure the identical mass flow. This procedure was done for all inlet nozzles, always with two nozzles of adjacent diameter. The result showed, that in any case the larger nozzle below a certain Re_D deviated from the result of the smaller nozzles. Interestingly the larger diameters yielded acceptable accuracy for $Re_D > 70000 - 80000$, while the smallest nozzles showed an unacceptable deviation even for $Re_D < 90000$. For low Re_D all nozzles showed a constant deviation of $\approx 5\%$ for $Re_D < 30.000$. In this Re-range the nozzles showed a constant offset to the calibration formula in DIN24163-2.

To verify the estimated reason of laminar flow at the pressure tap in the nozzle, the flow state in the nozzle has been examined. Therefore the 74mm-nozzle was equipped with a hot-film, which was placed between the circumferential pressure taps. Indeed the flow was found to be laminar for small flow rate, corresponding to the range of constant offset to the DIN-formula. Transition was obtained for $30000 < Re_D < 50000$ and for larger Re_D the hot-film indicated fully turbulent flow.

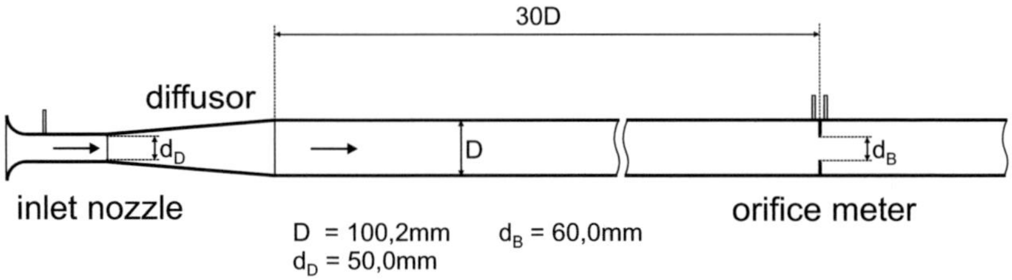

Figure A.1: Test rig for the calibration of inlet nozzles.

Since the orifice meter run enabled to install a inlet nozzle directly at the inlet of the pipe, the flow rate of both types of instruments could be reviewed simultaneously (compare Fig. A.1). The flow rate calculation based on the orifice meter is within the range of operation of the instrument and the thus the correct value, within the measurement accuracy of the orifice meter, is known. This calibration procedure has been conducted for the 50mm and 74mm inlet nozzles. Exemplary a comparison the determined flow rate from a simultaneous measurement with the 50mm nozzle and the 60mm orifice disc is shown in Figure A.2 (a). While the flow rate calculation with the orifice meter shows a linear relation on the rotation frequency of the ventilator, the result delivered by the inlet nozzle significantly deviates for small and moderate flow rate from a proportional relation. In Figure A.2 (b) the deviation is of the flow rate measurement with the nozzle from the measurement of the orifice is shown. If the Reynolds number in the nozzle Re_D is calculated, a flow rate of $200\text{m}^3/\text{h}$ approximately corresponds to the Re limit of $Re_D \geq 100.000$ given in DIN24153-2. Obviously below the Re-limit the formula according to the norm underestimates of the flow rate by approximately 2-3%. In the range of laminar flow at the flow rate is constantly overestimated by 5%.

From this data a correction function has been deduced. Since the curve of any approach did not accurately follow the complete data, the calibration function is split into the ranges: laminar, transitional and turbulent (below $Re_D = 100000$). Figure A.3 shows the results for the friction coefficient in dependence of Re_b for the 13% tripping in calculated with the DIN-formula (left) and corrected with the calibration function (right) on the downstream segment of the channel.

Although the calibration has been successful in the present case, there is one disadvantage in general application of the function. The transition in

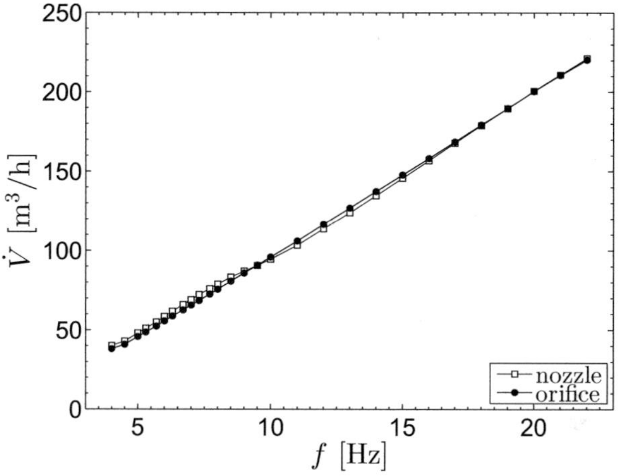

(a)

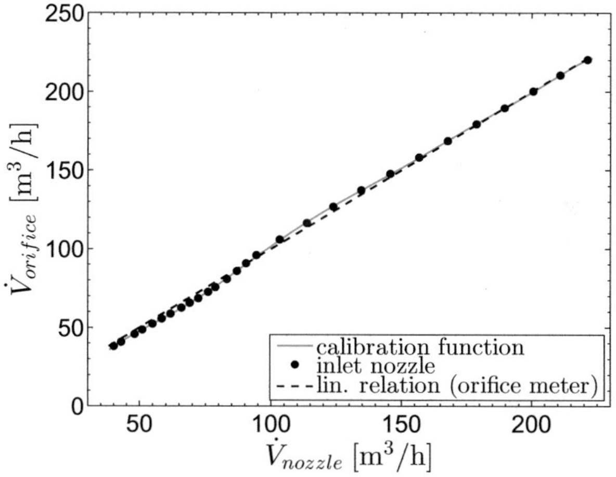

(b)

Figure A.2: Simultaneous measurement of the flow rate at varying motor frequency for the 50mm inlet nozzle the orifice meter with 60mm disc plate (a). Flow rate of the 50mm inlet nozzle over the flow rate of the orifice meter with 60mm disc plate (b).

the nozzle can be triggered by environmental noise and vibration. At the particular location of the nozzle in vicinity to the drive unit, this influence is dominated by the distortion originating from the blower and thus is always constant for the set-up. But in general application the nozzle, may not be located at a place, where the triggering is of defined magnitude and the transition becomes a function of the actual environmental conditions. From the experience with the inlet nozzles, it seems to be more promising to induce the transition with a tripping at the inlet of the nozzle and to calibrate the instrument for the extended turbulent range. It is important to carefully examine every little step in this procedure. Otherwise one can ruin the whole result, instead of gaining an extension of the measurement range.

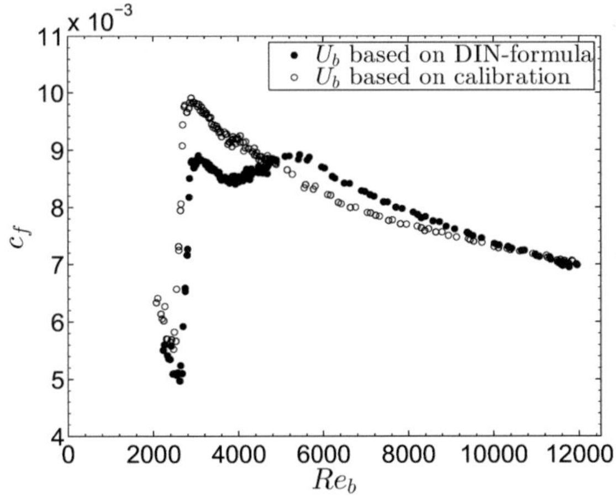

Figure A.3: Comparison of c_f with U_b calculated based on the DIN-formula and with the new calibration function.

A.2 Determination of the turbulence intensity without velocity calibration

In the following the derivation of the relation between the output voltage of the hot-wire anemometer and the turbulence level is explained. to the authors experience this relation delivers almost the same result as the determination from velocity calibrated data, which saves a lot of time.

King's law is often used for the calibration of hot-wire sensors

$$E(U) = \sqrt{A + BU^n}. \tag{A.1}$$

Once the constants A, B and n have been obtained in calibration by employing a least-square fitting procedure to data, the equation for the determination of the velocity reads

$$U(E) = \left(\frac{E^2 - A}{B}\right)^{\frac{1}{n}}. \tag{A.2}$$

In a fluctuating flow, the velocity and the voltage can be decomposed into a mean and superimposed disturbance part.

$$\begin{aligned} U &= \bar{U} + u' \\ E &= \bar{E} + e' \end{aligned} \tag{A.3}$$

A Taylor series expansion for the voltage $E(U)$ is used to linearize equation (A.1) by dropping all terms of higher order.

$$E(U) = E(\bar{U}) + \left(\frac{\partial E}{\partial U}\right)_{U=\bar{U}} (U - \bar{U}) + \ldots \tag{A.4}$$

The partial derivative

$$\left(\frac{\partial E}{\partial U}\right)_{U=\bar{U}} = \frac{1}{2}\frac{nB\bar{U}^{n-1}}{\sqrt{A + B\bar{U}^n}} \tag{A.5}$$

describes the gradient of the King's law at a particular velocity $U = \bar{U}$. As small velocity fluctuations are assumed for this derivation, the voltage fluctuation is closely described by the linear relation

$$e' = \left(\frac{\partial E}{\partial U}\right)_{U=\bar{U}} u'. \tag{A.6}$$

Rewriting this equation in terms of u'

$$u' = 2\frac{\sqrt{A + B\bar{U}^n}}{nB\bar{U}^{n-1}}\, e' = \frac{2\, e'\bar{E}}{nB\bar{U}^{n-1}} \tag{A.7}$$

and squaring the result

$$u'^2 = \frac{4\ e'^2 \bar{E}^2}{\left(nB\bar{U}^{n-1}\right)^2}, \tag{A.8}$$

one obtains the expression for the RMS value of the velocity fluctuation as a function of the the voltage RMS.

$$\sqrt{\overline{u'^2}} = \frac{2\bar{E}}{nB\bar{U}^{n-1}}\ \sqrt{\overline{e'^2}} \tag{A.9}$$

The one-dimensional turbulence intensity is defined as

$$\mathrm{Tu} = \frac{\sqrt{\overline{u'^2}}}{\bar{U}} = 2\frac{\sqrt{\overline{e'^2}}\ \bar{E}}{nB\bar{U}^{n-1}}\frac{1}{\bar{U}} = 2\frac{\sqrt{\overline{e'^2}}\ \bar{E}}{nB}\frac{1}{\bar{U}^n} \tag{A.10}$$

Substituting the relation

$$\bar{U}^n = \left(\frac{\bar{E}^2 - A}{B}\right), \tag{A.11}$$

the calibration constants B and n are eliminated from the problem

$$\mathrm{Tu} = \frac{\sqrt{\overline{u'^2}}}{\bar{U}} = 2\frac{\sqrt{\overline{e'^2}}\ \bar{E}}{nB}\frac{B}{\bar{E}^2 - A} \tag{A.12}$$

and the final expression for the final expression for the turbulence intensity is obtained.

$$\mathrm{Tu} = 2\frac{\sqrt{\overline{e'^2}}\bar{E}}{n\left(\bar{E}^2 - A\right)} \tag{A.13}$$

The calibration constant A can be approximated by measuring the mean voltage once at zero velocity $\bar{E}^2(U = 0) = A_0 \approx A$. The value A_0 is not exactly A since free convection is present in the measurement, which is not foreseen by the derivation of the original King's law. Nevertheless, the error introduced by this assumption is small and the Taylor series linearization yields very convenient and reasonably accurate results for low turbulence intensities.